Giacomo Lorenzoni

A New Law of Continuum Mechanics

https://www.giacomo.lorenzoni.name/
info@giacomo.lorenzoni.name; info@pec.giacomo.lorenzoni.name
ISBN 978-1-008-94549-4
I edition: June 2021

Contents

Introduction

Continuum Mechanics, physical science par excellence since it concerns the material reality commonly perceived by human senses, lacks its most important result i.e. a proven mathematical model of general applicability for the evolution of quantities describing mechanical and thermal aspects, devoid of errors implied by approximating laws (concepts exactly valid for all matter), and that is therefore affected only by errors, generally reducible but strictly speaking ineliminable, of relations that specify and characterize contingent substances.

In previous works ([77, 78, 79, 81]) we expounded (*pluralis modestiae*, as in the rest of this paper) a theory that, by translating fundamental concepts of Thermodynamics in local relations (i.e. between functions of points of a physical space) such as those of Continuum Mechanics and by introducing (mainly postulationally) completely new principles (such as the dynamic law object of this paper, the formulation of the first law of Thermodynamics, the expressions of total energy and friction heat), arrived at a system of PDEs (partial differential equations), of general applicability, thermomechanical due to the presence of quantities such as temperature, thermodynamic pressure, density and velocity, that can be solved since it consists of so many unknown functions in as many independent equations, devoid of approximations of laws, and therefore conceivable to constitute the aforementioned missing result.

However, regardless of the need to verify this new general model that may constitute future commitments, this work has the preparatory purpose of demonstrating, by means of deductions based on consolidated and shared foundations, the New Law (affirmed for the first time in [79] and expressed by (79)) that links density, velocity and thermodynamic pressure in the generic point of the material world. To this aim, the necessary tools of Logic, Tensor Calculus and Mathematical Analysis are widely provided, the conceptual representation of material reality is shown in detail (an Euclidean space, as well as the spacetime which is its four-dimensional specification with three volumic coordinates and one temporal, and by analogy also matter, are interpreted as discrete domains of definition of analytic functions, dividing them into infinite sets of infinitesimal elements to which are associated single numerical values), laws and principles of the state of the art are carefully detailed. In the context thus constituted, the argumentations leading to the New Law have been developed through systematic and coherent elaborations of original definitions and ideas consequent from careful reflections.

Lastly, after considering the current general model of the Continuous Mechanics, the best known models that specify it and the compatibility of these models with the New Law, significant improvements are expected from the new general model compared to the current one, as a consequence of the modeling errors of the material world peculiarly relevant and caused in the second by the constitutive equations that bind stress and strains.

Chapter 1

Preliminary notions

1.1 Logic

For the logic on which the following exposition is based, we refer specifically to sections 2 of [3] and 2.1 of [22].

A proposition is a sequence of graphic symbols, it may or may not have one or more meanings, and therefore may or may not be a further cognition than its mere graphic existence. Any accidental false meaning of a proposition is overlooked.

A cognition is qualified as implicit (i.e. we imply a cognition), to say that in the following it is valid even if not reaffirmed, but only if compatible as it is not contradicted. We name "default", or "of default", a setting predefined and implicit in the absence of specification.

An object is identified by the set of all its properties. A name is a proposition that communicates an object and represents it, and that attributes to it the properties indicated by its possible meanings. The names present in an argumentation relate the properties relatable, i.e. contingently relevant, indicated by them.

A quantity is an object reported by a name that gives it as a property a number that is named numeric value or only value. By default each number is positive and therefore the default value of each quantity is positive.

An A ≡ B states the following concepts relating to A and B: they are two names of the same object; they are equivalent in the sense of mutual substitutability so that one can replace the other, but only if the result is neither ambiguous nor erroneous as it could happen if invalidated by the impossible relationality between properties indicated by the meaning of the names. Therefore, to deduce a certain result from an A ≡ B it is necessary that in the inherent argumentation there is the condition constituted by being correctly relatable the properties indicated by the names. We attribute to "i.e." the same meaning as "≡".

Being $\mathcal{P}_{\rm A}$ and $\mathcal{P}_{\rm B}$ two propositions, we mean

$$\{\mathcal{P}_{\rm A} \parallel \mathcal{P}_{\rm B}\} \equiv \text{“}\mathcal{P}_{\rm A} \text{ subjected to the condition } \mathcal{P}_{\rm B}\text{”} \equiv \text{“}\mathcal{P}_{\rm A} \text{ of which } \mathcal{P}_{\rm B}\text{”} \equiv \text{“}\mathcal{P}_{\rm A} \text{ where } \mathcal{P}_{\rm B}\text{”}$$

and, with reference to (3) of [3],

$$\{\mathcal{P}_{\rm A} \Rightarrow \mathcal{P}_{\rm B}\} \equiv \{\mathcal{P}_{\rm B} \Leftarrow \mathcal{P}_{\rm A}\} \equiv \{\mathcal{P}_{\rm B}; \forall \mathcal{P}_{\rm A}\} \equiv \text{“from } \mathcal{P}_{\rm A} \text{ follows } \mathcal{P}_{\rm B}\text{”} \equiv \text{“}\mathcal{P}_{\rm A} \text{ implies } \mathcal{P}_{\rm B}\text{”}$$

The properties of the sets used in this paper are explained in section 2 of [3]. By meaning $\{\S_{\rm m}; {\rm m} = 1, \hat{\rm m}\} \equiv \{\S_1, \S_2, \dots \S_{\hat{\rm m}}\}$ (and $\{{\rm m} = 1, \hat{\rm m}\} \equiv \{{\rm m}; {\rm m} = 1, \hat{\rm m}\}$), a sequence and a set, both made up of $\hat{\rm m}$ elements, are respectively indicated $(\S_{\rm m}; {\rm m} = 1, \hat{\rm m})$ and $\{\S_{\rm m}; {\rm m} = 1, \hat{\rm m}\}$. A sequence is also a set and can therefore be treated as such.

We say that B is a specification of A to mean that B has all the compatible properties of A. We denote $\{\S \mid \mathcal{P}\}$ a set whose elements are all the different specifications of § contextually possible when exist the condition constituted by the proposition $\mathcal{P}$. We introduce

$$\text{Æ}\langle \underline{\rm A} \,/\!/\, {\rm B} \,/\!/\, {\rm C}\rangle \equiv \text{“}\{\S \parallel \S \in \underline{\rm A}\} \text{ is a specification of B of which C”}$$

where "$/\!/$ C" may be absent causing so the absence of "of which C".

We introduce

$$\text{“from: } {\rm A}_1; {\rm A}_2; \dots {\rm A}_{\hat{\rm i}}; \text{ follows } {\rm B}_0 \diamond_1 {\rm B}_1 \diamond_2 {\rm B}_2 \cdots \diamond_{\hat{\rm i}} {\rm B}_{\hat{\rm i}} \diamond_{\hat{\rm i}+1} {\rm B}_{\hat{\rm i}+1} \cdots \diamond_{\hat{\rm i}+\hat{\rm j}} {\rm B}_{\hat{\rm i}+\hat{\rm j}}\text{”} \equiv$$
$$\text{“}{\rm A}_1 \Rightarrow \{{\rm B}_0 \diamond_1 {\rm B}_1\}; {\rm A}_2 \Rightarrow \{{\rm B}_1 \diamond_2 {\rm B}_2\}; \dots {\rm A}_{\hat{\rm i}} \Rightarrow \{{\rm B}_{\hat{\rm i}-1} \diamond_{\hat{\rm i}} {\rm B}_{\hat{\rm i}}\}\text{”}$$

where: each of $\{\diamond_1, \diamond_2, \cdots \diamond_{\hat{\rm i}+\hat{\rm j}}\}$ is a generally different relational symbol, as for example one of $\{=, \neq, \equiv, \not\equiv\}$; $\{\diamond_{\hat{\rm i}+1} {\rm B}_{\hat{\rm i}+1} \cdots \diamond_{\hat{\rm i}+\hat{\rm j}} {\rm B}_{\hat{\rm i}+\hat{\rm j}}\}$ may be absent and if is present its validity is evident; each of $\{{\rm A}_1, {\rm A}_2, \dots {\rm A}_{\hat{\rm i}}\}$ is replaced by the symbol "þ" when the corresponding element of

$$\{\{{\rm B}_0 \diamond_1 {\rm B}_1\}, \{{\rm B}_1 \diamond_2 {\rm B}_2\}, \dots \{{\rm B}_{\hat{\rm i}-1} \diamond_{\hat{\rm i}} {\rm B}_{\hat{\rm i}}\}\}$$

is considered evident or its validity is highlighted after or constitutes a new definition.

We intend ${\rm A}\langle {\rm B}\rangle \equiv {\rm A}_{\rm B}$, $\wedge \equiv$ AND $\equiv$ "conjunction", $\veebar \equiv$ XOR $\equiv$ "exclusive disjunction".

1.2 Tensors in an Euclidean space

An infinite set is constituted by an unlimitedly large number of elements. A direction is the evident common property of an infinite set of parallel straight lines. A direction and one of its two senses constitute an oriented direction. A straight line is oriented inasmuch is defined the sense in which an abscissa increases. An oriented straight line has the its own evident oriented direction, which is the same for an infinite set of parallel and equally oriented straight lines.

A point is a geometric object without extent, that does not take up space and cannot be divided into parts. Consistently with this, lines and surfaces have no thickness. An $\hat{\mathrm{i}}$-dimensional Euclidean space $\mathcal{E}_{\hat{\mathrm{i}}}$ is an infinite set of adjacent points in biunivocal (one-to-one) correspondence with the set of each different $\hat{\mathrm{i}}$-tuple (i.e. sequence of $\hat{\mathrm{i}}$ elements) of real numbers. This adjacency will be specified as minimum distances unlimitedly small, and concerns lines and surfaces analogously to respective $\mathcal{E}_1$ and $\mathcal{E}_2$.

An $\mathcal{E}_{\hat{\mathrm{i}}}$ has an orthogonal Cartesian reference system whose $\hat{\mathrm{i}}$ coordinate axes are mutually orthogonal straight lines, oriented by measuring on them the respective coordinates $\underline{\mathrm{x}}$ of which $\underline{\mathrm{x}} \equiv \left(\mathrm{x}_{\mathrm{i}}; \mathrm{i} = 1, \hat{\mathrm{i}}\right)$ and $-\infty < \mathrm{x}_{\mathrm{i}} < \infty$, and that intersect at the origin point where all $\underline{\mathrm{x}}$ have null value. On the basis of said biunivocal correspondence, the name $\underline{\mathrm{x}}$ is also used to refer to the corresponding point. The i-th direction is the oriented direction of the i-th coordinate axis.

A tensor $\mathbf{A}$, of order o and inherent to $\mathcal{E}_{\hat{\mathrm{i}}}$, consists of $\hat{\mathrm{i}}^{\mathsf{o}}$ components, the generic of which is indicated $A_{\mathrm{i}_1\mathrm{i}_2\ldots\mathrm{i}_{\mathsf{o}}}$ of which $\mathrm{i}_o \in \left\{\mathrm{i} = 1, \hat{\mathrm{i}}\right\}$ with $o \in \{\S = 1, \mathsf{o}\}$. An $A_{\mathrm{i}_1\mathrm{i}_2\ldots\mathrm{i}_{\mathsf{o}}}$ is a scalar, i.e. a real number, associated with the o-tuple constituted by the particular o coordinate directions indicated by $(\mathrm{i}_1, \mathrm{i}_2, \ldots \mathrm{i}_{\mathsf{o}})$, being this o-tuple a sequence of o elements chosen arbitrarily among those of $\left\{\mathrm{i} = 1, \hat{\mathrm{i}}\right\}$ and therefore being a disposition with repetition of class o of $\hat{\mathrm{i}}$ objects. For such a tensor, besides the direct notation $\mathbf{A}$, we also have the indicial notation $A_{\mathrm{i}_1\mathrm{i}_2\ldots\mathrm{i}_{\mathsf{o}}}$ in the sense of

$$\mathbf{A} \equiv A_{\mathrm{i}_1\mathrm{i}_2\ldots\mathrm{i}_{\mathsf{o}}} \tag{1}$$

where any quantity present in the name $\mathbf{A}$ is treated as a constant with respect to the $(\mathrm{i}_1, \mathrm{i}_2, \ldots \mathrm{i}_{\mathsf{o}})$ and an eventual index in denominator is considered more to the right than one in numerator.

Naming **B** a tensor of which $\mathbf{B} \equiv B_{\mathrm{m_1 m_2 \ldots m_{\mathfrak{p}}}}$, the addition of **A** and **B** is indicated **A**+**B** and, only if $\mathfrak{p} = \mathfrak{o}$, is equivalent to the tensor $\mathbf{C}_\mathrm{S}$ as it is shown by

$$\mathbf{A} + \mathbf{B} \equiv A_{\mathrm{i_1 i_2 \ldots i_{\mathfrak{o}}}} + B_{\mathrm{i_1 i_2 \ldots i_{\mathfrak{o}}}} \equiv C_{\mathrm{S i_1 i_2 \ldots i_{\mathfrak{o}}}} \equiv \mathbf{C}_\mathrm{S} \tag{2}$$

The direct (or tensor or outer) product of **A** and **B** is indicated **AB** and is equivalent to the tensor $\mathbf{C}_\mathrm{D}$ as it is shown by

$$\mathbf{AB} \equiv \mathbf{A} \otimes \mathbf{B} \equiv A_{\mathrm{i_1 i_2 \ldots i_{\mathfrak{o}}}} B_{\mathrm{m_1 m_2 \ldots m_{\mathfrak{p}}}} \equiv C_{\mathrm{D i_1 i_2 \ldots i_{\mathfrak{o}} m_1 m_2 \ldots m_{\mathfrak{p}}}} \equiv \mathbf{C}_\mathrm{D} \tag{3}$$

The inner product of **A** and **B** is indicated $\langle \mathbf{A}, \mathbf{B} \rangle$, and, only if $\mathfrak{o} > 0$ and $\mathfrak{p} > 0$, is equivalent to the tensor $\mathbf{C}_\mathrm{I}$ as it is shown by

$$\langle \mathbf{A}, \mathbf{B} \rangle \equiv \sum_{\mathrm{i}=1}^{\hat{\mathrm{i}}} A_{\mathrm{i_1 i_2 \ldots i_{\mathfrak{o}-1} i}} B_{\mathrm{i m_2 m_3 \ldots m_{\mathfrak{p}}}} \equiv C_{\mathrm{I i_1 i_2 \ldots i_{\mathfrak{o}-1} m_2 m_3 \ldots m_{\mathfrak{p}}}} \equiv \mathbf{C}_\mathrm{I} \tag{4}$$

of which $\langle \mathbf{A}, \mathbf{B} \rangle \equiv \mathbf{A} \cdot \mathbf{B}$ that, as we'll see, is introduced in coherence with (14). An $\langle \mathbf{A}, \mathbf{B} \rangle$ is defined placing equal to a same index (different from those already present) the two closest indices between those of $A_{\mathrm{i_1 i_2 \ldots i_{\mathfrak{o}}}}$ and $B_{\mathrm{m_1 m_2 \ldots m_{\mathfrak{p}}}}$ and then adding the $\hat{\mathrm{i}}$ addends corresponding to the just as many $\hat{\mathrm{i}}$ values of said same index.

In an expression, the direct products are considered first, then the inner products and then the sums.

A vector is a straight segment defined by its oriented direction (by a direction inasmuch it can lie on any element of its infinite set of parallel straight lines, and by one of the two senses of this direction inasmuch its extreme points are distinct as origin and destination i.e. as initial and final) and by its length i.e. magnitude. Being so the magnitude of a vector a non-negative quantity, it is indicated by the name of this vector between delimiters equal to "|", e.g. the magnitude of a vector **a** is indicated $|\mathbf{a}|$ of which $|\mathbf{a}| \geq 0$.

In the reference system of $\mathcal{E}_{\hat{\mathrm{i}}}$, an oriented direction is identified by $\hat{\mathrm{i}}$ direction cosines, that are the same for each element of an infinite set of parallel oriented straight lines, the i-th of which is the cosine of the convex angle between such an element and the i-th coordinate axis, being this angle that of a rotate, less than 180°, of one or the other in their common plane until they both have the same oriented direction.

Moreover, naming $\left(\hat{\mathbf{e}}_\mathrm{i}; \mathrm{i} = 1, \hat{\mathrm{i}}\right)$ the direction cosines that identify the oriented direction of **a**, for the known property of a right triangle, we have

$\mathbf{a}_{\mathrm{i}} = |\mathbf{a}|\hat{\mathbf{e}}_{\mathrm{i}}$ whose $\mathbf{a}_{\mathrm{i}}$ verifies the expression $|\mathbf{a}| = \sqrt{\sum_{\mathrm{i}=1}^{\hat{\mathrm{i}}} \mathbf{a}_{\mathrm{i}}^2}$ of an Euclidean distance. Therefore $\mathbf{a}$ is in biunivocal correspondence with the real numbers $(\mathbf{a}_{\mathrm{i}}; \mathrm{i} = 1, \hat{\mathrm{i}})$, since magnitude and oriented direction of $\mathbf{a}$ allow to determine $\mathbf{a}_{\mathrm{i}}$ by means of $\mathbf{a}_{\mathrm{i}} = |\mathbf{a}|\hat{\mathbf{e}}_{\mathrm{i}}$, and *vice versa* the $(\mathbf{a}_{\mathrm{i}}; \mathrm{i} = 1, \hat{\mathrm{i}})$ allow to determine $\mathbf{a}$ by means of

$$|\mathbf{a}| = \sqrt{\sum_{\mathrm{i}=1}^{\hat{\mathrm{i}}} \mathbf{a}_{\mathrm{i}}^2} \qquad \hat{\mathbf{e}}_{\mathrm{i}} = \frac{\mathbf{a}_{\mathrm{i}}}{|\mathbf{a}|} \tag{5}$$

On the basis of this biunivocal correspondence we introduce $\mathbf{a} \equiv \mathbf{a}_{\mathrm{i}}$ whereby a vector is a tensor of order 1.

A versor is a vector that has unitary magnitude. Naming $\hat{\mathbf{e}}$ the versor that has the same oriented direction as $\mathbf{a}$, we have $|\hat{\mathbf{e}}| = 1$ and the direction cosines of $\hat{\mathbf{e}}$ are the same $(\hat{\mathbf{e}}_{\mathrm{i}}; \mathrm{i} = 1, \hat{\mathrm{i}})$ of $\mathbf{a}$. This and the being the i-th direction cosine of a vector equal to the ratio of its i-th component to its magnitude give rise to $\hat{\mathbf{e}} \equiv \hat{\mathbf{e}}_{\mathrm{i}}$, by following that the i-th direction cosine of a versor is also its i-th component.

This, naming $\hat{\boldsymbol{e}}_{\mathrm{i}}$ the versor that has same oriented direction of the i-th coordinate axis (whereby the $(\hat{\boldsymbol{e}}_{\mathrm{i}}; \mathrm{i} = 1, \hat{\mathrm{i}})$ are the coordinate versors of the reference system) and being the direction cosines of such axis the $(\delta_{\mathrm{i}i}; i = 1, \hat{\mathrm{i}})$ defined by

$$\{\delta_{\mathrm{mn}} = 1; \forall \mathrm{m} = \mathrm{n}\} \qquad \{\delta_{\mathrm{mn}} = 0; \forall \mathrm{m} \neq \mathrm{n}\} \tag{6}$$

with δ_{mn} the Kronecker delta, they bring $\hat{\boldsymbol{e}}_{\mathrm{i}} \equiv \delta_{\mathrm{i}i}$.

Based on (6) and

$$\text{Æ}\langle(\mathbf{a}_i, \mathbf{a}), (\delta_{\mathrm{ii}}, \hat{\boldsymbol{e}}_i) \,/\!/\, (A_{\mathrm{i}_1\mathrm{i}_2\ldots\mathrm{i}_{\mathrm{o}}}, \mathbf{A}) \,/\!/\, (1)\rangle$$

we have

$$\mathbf{a}_{\mathrm{i}} = \sum_{i=1}^{\hat{\mathrm{i}}} \mathbf{a}_i \delta_{\mathrm{i}i} \qquad \mathbf{a} = \sum_{\mathrm{i}=1}^{\hat{\mathrm{i}}} \mathbf{a}_{\mathrm{i}} \hat{\boldsymbol{e}}_{\mathrm{i}} \tag{7}$$

as well as, for o=2,

$$A_{\mathrm{i}i} = \sum_{m=1}^{\hat{\mathrm{i}}} \sum_{\mathrm{m}=1}^{\hat{\mathrm{i}}} A_{\mathrm{m}m} \delta_{\mathrm{mi}} \delta_{mi} \qquad \mathbf{A} = \sum_{i=1}^{\hat{\mathrm{i}}} \sum_{\mathrm{i}=1}^{\hat{\mathrm{i}}} A_{\mathrm{i}i} \hat{\boldsymbol{e}}_{\mathrm{i}} \hat{\boldsymbol{e}}_i$$

From (7) follows

$$-\mathbf{a} = \sum_{i=1}^{\hat{i}} -\mathbf{a}_i \hat{\boldsymbol{e}}_i$$

that, for

$$Æ\langle -\mathbf{a}, -\mathbf{a}_i \,/\!/\, \mathbf{a}, \mathbf{a}_i \,/\!/\, (7)\rangle,$$

gives rise to $-\mathbf{a} \equiv -\mathbf{a}_i$ from which it is deduced that $-\mathbf{a}$ has, with respect to $\mathbf{a}$, equal magnitude and direction, and opposite sense. For this reason $\mathbf{a}$ and $-\mathbf{a}$ are called equal and opposite.

From: (7);

$$|\mathbf{a}|^{-1}\mathbf{a}_i = \hat{\mathbf{e}}_i; \qquad Æ\langle \hat{\mathbf{e}}, \hat{\mathbf{e}}_i \,/\!/\, \mathbf{a}, \mathbf{a}_i \,/\!/\, (7)\rangle;$$

follows

$$|\mathbf{a}|^{-1}\mathbf{a} = \sum_{i=1}^{\hat{i}} |\mathbf{a}|^{-1}\mathbf{a}_i \hat{\boldsymbol{e}}_i = \sum_{i=1}^{\hat{i}} \hat{\mathbf{e}}_i \hat{\boldsymbol{e}}_i = \hat{\mathbf{e}}$$

that shows $\mathbf{a} = |\mathbf{a}|\hat{\mathbf{e}}$. By placing

$$\left\{ \varpi_{§} = \frac{§}{|§|}; \forall § \neq 0 \right\} \qquad \{\varpi_{§} = 0; \forall § = 0\} \tag{8}$$

and meaning $\mathbf{a}$ a scalar and $\hat{\mathbf{e}}_{\mathbf{a}}$ a versor of which $\hat{\mathbf{e}}_{\mathbf{a}} \equiv \hat{\mathbf{e}}_{\mathbf{a}i}$, we have

$$\mathbf{a}\hat{\mathbf{e}}_{\mathbf{a}} = |\mathbf{a}|\varpi_{\mathbf{a}}\hat{\mathbf{e}}_{\mathbf{a}} = |\mathbf{a}|\hat{\boldsymbol{e}}_{\mathbf{a}} \tag{9}$$

of which

$$\hat{\boldsymbol{e}}_{\mathbf{a}} = \varpi_{\mathbf{a}}\hat{\mathbf{e}}_{\mathbf{a}} \qquad \text{i.e.} \qquad \hat{\boldsymbol{e}}_{\mathbf{a}} \equiv \hat{\mathbf{e}}_{\mathbf{a}} \veebar -\hat{\mathbf{e}}_{\mathbf{a}}$$

and therefore $|\hat{\boldsymbol{e}}_{\mathbf{a}}| = 1$.

This and (9) show

$$Æ\langle \mathbf{a}\hat{\mathbf{e}}_{\mathbf{a}}, |\mathbf{a}|\hat{\boldsymbol{e}}_{\mathbf{a}} \,/\!/\, \mathbf{a}, |\mathbf{a}|\hat{\mathbf{e}}\rangle$$

so the vector $\mathbf{a}\hat{\mathbf{e}}_{\mathbf{a}}$ has magnitude $|\mathbf{a}|$ and magnitude with sign $\mathbf{a}$.

Indeed, from:

$$\text{Æ}\langle \mathbf{a\hat{e}_a}, \mathbf{a\hat{e}_{ai}} \,/\!/\, \mathbf{a}, \mathbf{a}_\mathrm{i} \,/\!/\, (5)\rangle;$$

$$\sum_{i=1}^{\hat{i}} \hat{\mathbf{e}}_i^2 = 1 \text{ which is deduced from (5);}$$

follows

$$|\mathbf{a\hat{e}_a}| = |\mathbf{a}|\sqrt{\sum_{i=1}^{\hat{i}} \hat{\mathbf{e}}_{\mathbf{a}i}^2} = |\mathbf{a}| \tag{10}$$

A vector is applied if is specified its application point i.e. its origin. The scalar (or dot) product of the vectors **a** and **b**, of which $\mathbf{b} \equiv \mathbf{b}_\mathrm{i}$, is the scalar indicated $\mathbf{a} \cdot \mathbf{b}$ and defined by

$$\mathbf{a} \cdot \mathbf{b} \equiv |\mathbf{a}||\mathbf{b}| \cos\theta \tag{11}$$

where θ is the angle (less than 180°) between **a** and **b** when are applied to a same point. This product is commutative (i.e. $\mathbf{a} \cdot \mathbf{b} = \mathbf{b} \cdot \mathbf{a}$) and distributive with respect to the sum.

The projection of **a** on a versor $\hat{\mathbf{e}}$ (i.e. on the oriented direction of a such $\hat{\mathbf{e}}$) is defined $(\hat{\mathbf{e}} \cdot \mathbf{a})\hat{\mathbf{e}}$ and then has magnitude with sign $\hat{\mathbf{e}} \cdot \mathbf{a}$.

Two versors $\hat{\mathbf{e}}_\mathrm{N}$ and $\hat{\mathbf{e}}_\mathrm{T}$, of which $\hat{\mathbf{e}}_\mathrm{N} \cdot \hat{\mathbf{e}}_\mathrm{T} = 0$, are those coordinate of a two-dimensional reference system where **a**, if lies on same plane of $\hat{\mathbf{e}}_\mathrm{N}$ and $\hat{\mathbf{e}}_\mathrm{T}$, is expressed, analogously to (7), by

$$\mathbf{a} = (\hat{\mathbf{e}}_\mathrm{N} \cdot \mathbf{a})\hat{\mathbf{e}}_\mathrm{N} + (\hat{\mathbf{e}}_\mathrm{T} \cdot \mathbf{a})\hat{\mathbf{e}}_\mathrm{T} \tag{12}$$

whose $\hat{\mathbf{e}}_\mathrm{N} \cdot \mathbf{a}$ and $\hat{\mathbf{e}}_\mathrm{T} \cdot \mathbf{a}$ are the components of **a** along $\hat{\mathbf{e}}_\mathrm{N}$ and $\hat{\mathbf{e}}_\mathrm{T}$, as $\mathbf{a}_\mathrm{i}$ is the component of **a** along $\mathrm{x_i}$ i.e. along $\hat{e}_\mathrm{i}$.

From:

$$\text{Æ}\langle \mathbf{a}\hat{\boldsymbol{e}}_\mathrm{i}, \mathbf{b}\hat{\boldsymbol{e}}_i, \theta_{\mathrm{i}i} \,/\!/\, \mathbf{a}, \mathbf{b}, \theta \,/\!/\, (11)\rangle;$$

$$\text{Æ}\langle \mathbf{a}\hat{\boldsymbol{e}}_\mathrm{i}, \mathbf{b}\hat{\boldsymbol{e}}_i \,/\!/\, \mathbf{a\hat{e}_a} \,/\!/\, (10)\rangle, \qquad \varpi_{\S} = |\S|/\S;$$

$$\varpi_\mathbf{a}\varpi_\mathbf{b} \cos\theta_{\mathrm{i}i} = \delta_{\mathrm{i}i};$$

follows

$$\mathbf{a}\hat{\boldsymbol{e}}_\mathrm{i} \cdot \mathbf{b}\hat{\boldsymbol{e}}_i \equiv |\mathbf{a}\hat{\boldsymbol{e}}_\mathrm{i}||\mathbf{b}\hat{\boldsymbol{e}}_i| \cos\theta_{\mathrm{i}i} = \mathbf{ab}\varpi_\mathbf{a}\varpi_\mathbf{b} \cos\theta_{\mathrm{i}i} = \mathbf{ab}\delta_{\mathrm{i}i} \tag{13}$$

From: (7); distributivity of scalar product with respect to sum; (13); (4); follows

$$\mathbf{a}\cdot\mathbf{b}=\left(\sum_{\mathrm{i}=1}^{\hat{\mathrm{i}}}\mathbf{a}_{\mathrm{i}}\hat{\boldsymbol{e}}_{\mathrm{i}}\right)\cdot\left(\sum_{i=1}^{\hat{\mathrm{i}}}\mathbf{b}_{i}\hat{\boldsymbol{e}}_{i}\right)=\sum_{\mathrm{i}=1}^{\hat{\mathrm{i}}}\sum_{i=1}^{\hat{\mathrm{i}}}\mathbf{a}_{\mathrm{i}}\hat{\boldsymbol{e}}_{\mathrm{i}}\cdot\mathbf{b}_{i}\hat{\boldsymbol{e}}_{i}=\sum_{\mathrm{i}=1}^{\hat{\mathrm{i}}}\mathbf{a}_{\mathrm{i}}\mathbf{b}_{\mathrm{i}}=\langle\mathbf{a},\mathbf{b}\rangle \tag{14}$$

whereby we introduce $\mathbf{A}\cdot\mathbf{B}\equiv\langle\mathbf{A},\mathbf{B}\rangle$ inherent to (4).

The direct product **ab** is a tensor of order 2 named dyad or dyadic product or tensorial product.

The gradient and divergence of $\mathbf{A}$ are expressed by $\nabla\mathbf{A}$ and $\nabla\cdot\mathbf{A}$, whose operator ∇ (named "nabla" or "del") is the symbolic vector defined by

$$\nabla\equiv\frac{\partial}{\partial \mathrm{x_i}}\qquad\text{of which}\qquad\left(\frac{\partial}{\partial \mathrm{x_i}}\right)\c{s}\equiv\frac{\partial\c{s}}{\partial \mathrm{x_i}}.$$

For $\mathrm{o}=0$ and so $\mathbf{A}\equiv A$,

$$\nabla\mathbf{A}\equiv\frac{\partial A}{\partial \mathrm{x_i}}\qquad\text{and}\qquad\nabla\cdot\mathbf{A}\text{ is not defined;}$$

for $\mathrm{o}=1$,

$$\nabla\mathbf{A}\equiv\frac{\partial A_i}{\partial \mathrm{x_i}}\qquad\text{and}\qquad\nabla\cdot\mathbf{A}\equiv\sum_{\mathrm{i}=1}^{\hat{\mathrm{i}}}\frac{\partial A_{\mathrm{i}}}{\partial \mathrm{x_i}};$$

for $\mathrm{o}=2$,

$$\nabla\mathbf{A}\equiv\frac{\partial A_{\mathrm{i_2 i_3}}}{\partial \mathrm{x_{i_1}}}\qquad\text{and}\qquad\nabla\cdot\mathbf{A}\equiv\sum_{\mathrm{i}=1}^{\hat{\mathrm{i}}}\frac{\partial A_{\mathrm{i}i}}{\partial \mathrm{x_i}}.$$

Thus $\nabla\mathbf{A}$ and $\nabla\cdot\mathbf{A}$ are two tensors whose order is respectively greater and less than 1 with respect to that of $\mathbf{A}$.

In relation to (2) we have

$$\nabla(\mathbf{A}+\mathbf{B})\equiv\frac{\partial(A_{\mathrm{i_1 i_2\ldots i_o}}+B_{\mathrm{i_1 i_2\ldots i_o}})}{\partial \mathrm{x_i}}=\frac{\partial A_{\mathrm{i_1 i_2\ldots i_o}}}{\partial \mathrm{x_i}}+\frac{\partial B_{\mathrm{i_1 i_2\ldots i_o}}}{\partial \mathrm{x_i}}\equiv\nabla\mathbf{A}+\nabla\mathbf{B}$$

$$\nabla\cdot\mathbf{A}+\nabla\cdot\mathbf{B}\equiv\sum_{\mathrm{i}=1}^{\hat{\mathrm{i}}}\frac{\partial A_{\mathrm{ii_2\ldots i_o}}}{\partial \mathrm{x_i}}+\frac{\partial B_{\mathrm{ii_2\ldots i_o}}}{\partial \mathrm{x_i}}=\sum_{\mathrm{i}=1}^{\hat{\mathrm{i}}}\frac{\partial C_{\mathrm{Sii_2\ldots i_o}}}{\partial \mathrm{x_i}}\equiv \nabla\cdot\mathbf{C}_{\mathrm{S}}\equiv\nabla\cdot(\mathbf{A}+\mathbf{B}) \tag{15}$$

In relation to (3) we deduce

$$\nabla(\mathbf{AB}) \equiv \frac{\partial A_{\mathrm{i}_1\mathrm{i}_2\ldots\mathrm{i}_o} B_{\mathrm{m}_1\mathrm{m}_2\ldots\mathrm{m}_p}}{\partial \mathrm{x}_\mathrm{i}} =$$

$$A_{\mathrm{i}_1\mathrm{i}_2\ldots\mathrm{i}_o} \frac{\partial B_{\mathrm{m}_1\mathrm{m}_2\ldots\mathrm{m}_p}}{\partial \mathrm{x}_\mathrm{i}} + B_{\mathrm{m}_1\mathrm{m}_2\ldots\mathrm{m}_p} \frac{\partial A_{\mathrm{i}_1\mathrm{i}_2\ldots\mathrm{i}_o}}{\partial \mathrm{x}_\mathrm{i}} \equiv \mathbf{A}\nabla\mathbf{B} + \mathbf{B}\nabla\mathbf{A}$$

$$\nabla \cdot (\mathbf{AB}) \equiv \nabla \cdot \mathbf{C}_\mathrm{D} \equiv \sum_{\mathrm{i}=1}^{\hat{\mathrm{i}}} \frac{\partial C_{\mathrm{Dii}_2\ldots\mathrm{i}_o\mathrm{m}_1\mathrm{m}_2\ldots\mathrm{m}_p}}{\partial \mathrm{x}_\mathrm{i}} \equiv \sum_{\mathrm{i}=1}^{\hat{\mathrm{i}}} \frac{\partial A_{\mathrm{ii}_2\ldots\mathrm{i}_o} B_{\mathrm{m}_1\mathrm{m}_2\ldots\mathrm{m}_p}}{\partial \mathrm{x}_\mathrm{i}} =$$

$$\sum_{\mathrm{i}=1}^{\hat{\mathrm{i}}} A_{\mathrm{ii}_2\ldots\mathrm{i}_o} \frac{\partial B_{\mathrm{m}_1\mathrm{m}_2\ldots\mathrm{m}_p}}{\partial \mathrm{x}_\mathrm{i}} + B_{\mathrm{m}_1\mathrm{m}_2\ldots\mathrm{m}_p} \frac{\partial A_{\mathrm{ii}_2\ldots\mathrm{i}_o}}{\partial \mathrm{x}_\mathrm{i}} \equiv \nabla\mathbf{B} \cdot \mathbf{A} + \mathbf{B}\nabla \cdot \mathbf{A}$$

and, being A a scalar,

$$\begin{aligned} \nabla \cdot \mathbf{B}A = \nabla \cdot A\mathbf{B} \equiv \sum_{\mathrm{i}=1}^{\hat{\mathrm{i}}} \frac{\partial A B_{\mathrm{im}_2\ldots\mathrm{m}_p}}{\partial \mathrm{x}_\mathrm{i}} = \\ A \sum_{\mathrm{i}=1}^{\hat{\mathrm{i}}} \frac{\partial B_{\mathrm{im}_2\ldots\mathrm{m}_p}}{\partial \mathrm{x}_\mathrm{i}} + \sum_{\mathrm{i}=1}^{\hat{\mathrm{i}}} B_{\mathrm{im}_2\ldots\mathrm{m}_p} \frac{\partial A}{\partial \mathrm{x}_\mathrm{i}} \equiv A\nabla \cdot \mathbf{B} + \nabla A \cdot \mathbf{B} \end{aligned} \quad (16)$$

1.3 Mathematical Analysis

A quantity is a property of an object because describes it specifically by means of its properties, and in particular is a scalar if these properties consist only in the numeric value or is an applied vector if such properties are magnitude, direction, sense and application point.

Is named $\mathbb{R}\langle \mathrm{y} \rangle$ the set of different values the quantity y can have. Such an y is a constant or a variable respectively if $\mathbb{R}\langle \mathrm{y} \rangle$ is constituted by one or more elements.

To the $\hat{\mathrm{m}}$ quantities $\underline{\mathrm{y}}$, of which $\underline{\mathrm{y}} \equiv (\mathrm{y}_\mathrm{m}; \mathrm{m} = 1, \hat{\mathrm{m}})$, is associated the set $\mathbb{R}\langle \underline{\mathrm{y}} \rangle$ of all different $\hat{\mathrm{m}}$-tuples of possible values of such quantities.

We use a same name for a part of $\boldsymbol{\mathcal{E}}_{\hat{\mathrm{i}}}$ and a measure (of the extent, as is implicit) of such part.

1.3.1 Functions, limits and derivatives.

An analytic function $f(\underline{\mathrm{x}})$ is a mathematical formula, whose quantities are all treated as constants except its independent variables $\underline{\mathrm{x}}$, and that in correspondence of each $\hat{\text{i}}$-tuple of values of the $\underline{x}$ has one o more numerical values in the respective cases that is monodrome (as is implicit) or polydrome.

The domain of definition of $f(\underline{x})$ is $\Im\langle f(\underline{x})\rangle$ and is constituted by every different $\hat{\text{i}}$-tuple of values of the $\underline{x}$ to which corresponds univocally the inherent value of $f(\underline{x})$. The $\underline{x}$ are said independent in consequence of their reciprocal independence, that is indicated with $\ddot{\text{I}}\langle \underline{x}\rangle$ to mean that each can have every its own value present in $\Im\langle f(\underline{x})\rangle$ regardless of which particular values the remaining $\hat{\text{i}}-1$ have contingently. The substitution in $f(\underline{x})$ of some of the $\underline{x}$ with other variables (or functions, in which case we have a composite function) implies generally a different domain of definition.

The limit of A as B approaches C (such an approach is indicated $\mathsf{B} \to \mathsf{C}$) is the object to which A gets closer and closer when B gets closer and closer to C subordinately to the condition $\mathsf{B} \not\equiv \mathsf{C}$, is indicated $\lim_{\mathsf{B}\to\mathsf{C}} \mathsf{A}$, is defined if to each B corresponds an only A. Therefore a limit is only conceptual and not exists in the real material world because consists in an unlimited becoming.

Coherently with this, the limit of $f(\underline{x})$ as $\underline{x}$ approaches the point $\underline{x}_0$ of $\boldsymbol{\mathcal{E}}_{\hat{\text{i}}}$ have the fundamental analytic definition (e.g. (2.4.2.6) of [22]) according to which a

$$\lim_{\underline{x}\to\underline{x}_0} f(\underline{x}) = \text{L}$$

is equivalent to

$$\{|f(\underline{x}) - \text{L}| < \varepsilon; \forall \varepsilon > 0, \forall \{\underline{x} \parallel \underline{x} \not\equiv \underline{x}_0\} \in \mathcal{I} \cap \Im\langle f(\underline{x})\rangle\}$$

where $\mathcal{I}$ is a neighborhood of $\underline{x}_0$.

With reference to the known concepts inherent the quantities limitlessly small and large named infinitesimals and infinites (and to the duality of such denominations), is implicit that: an infinitesimal of a certain order is not such in absolute, but only because is related to an infinitesimal of order 0 constituted by a quantity finite i.e. limited; an infinitesimal is of order 1; infinitesimals have same order; an object is named infinitesimal inasmuch is so one of its measures; an infinitesimal of higher order, i.e. greater than 1, is negligible; if $\mathbf{a}$ and $\mathbf{b}$ are infinitesimals of order $O_{\mathbf{a}}$ and $O_{\mathbf{b}}$, $\mathbf{a}$ is an infinitesimal of order $O_{\mathbf{a}} - O_{\mathbf{b}}$ compared to $\mathbf{b}$.

A $f(\underline{x}) \in \mathsf{C}^0(\underline{x}_0)$ affirms the continuity of $f(\underline{x})$ at $\underline{x}_0$ and has the expression

$$\{f(\underline{x}) \in \mathsf{C}^0(\underline{x}_0)\} \equiv \left\{\lim_{\underline{x}\to\underline{x}_0} f(\underline{x}) = f(\underline{x}_0) \neq \pm\infty\right\} \tag{17}$$

whose second member is coherent with the being $f(\underline{x}) - f(\underline{x}_0)$ infinitesimal if $\underline{x} \to \underline{x}_0$.

Each function is implicitly continuous in every point of its definition domain, because we suppose that every point of non-continuity is eliminable by mean of usual techniques of Mathematical Analysis. Being so implicit, for each function, the specification of second member of (17), we have

$$\{f_A(\underline{a}) = f_B(\underline{b})\} \equiv \left\{\lim_{\underline{a}\to\underline{a}} f_A(\underline{a}) = \lim_{\underline{b}\to\underline{b}} f_B(\underline{b})\right\} \tag{18}$$

without which $f_A(\underline{a}) = f_B(\underline{b})$ is valid for $f_A(\underline{a})$ and $f_B(\underline{b})$ understood as two values deduced from the only two points $\underline{a}$ and $\underline{b}$, and not deduced instead from the definition of a limit i.e. excluding exactly such two values and using instead those corresponding to the two infinite sets of points respectively implied by the neighborhoods of $\underline{a}$ and $\underline{b}$.

A $y = f(\underline{x})$ expresses the quantity y as function of $\underline{x}$ inasmuch attributes the value of $f(\underline{x})$ to the dependent variable y, is equivalent to saying that y is expressed by $f(\underline{x})$, and it is enough for say that y is function of $\underline{x}$. A $\underline{y}(\underline{x})$ implies $\underline{y} = \underline{y}(\underline{x})$ of which

$$\text{“}\underline{y} = \underline{y}(\underline{x})\text{”} \equiv (y_m = y_m(\underline{x}); m = 1, \hat{m}),$$

and so $f(\underline{x})$ implies $f = f(\underline{x})$. A $f(\underline{x})$ is named constant if has one same value anyway vary the independent variables.

The Δy, δy and the total differential dy are three differences between two values of y i.e. variations of this quantity, but Δy is a quantity finite whereas δy and dy are infinitesimals inasmuch defined by $\delta y = \lim_{\Delta y \to 0} \Delta y$ and, only if $y = f(\underline{x})$, by

$$dy \equiv df(\underline{x}) \equiv \sum_{i=1}^{\hat{i}} \frac{\partial f(\underline{x})}{\partial x_i} dx_i \tag{19}$$

of which $dx_i \equiv \delta x_i$ and where $\partial f(\underline{x})/\partial x_i$ is the partial derivative, of $f(\underline{x})$ with respect to x_i, in turn defined, as limit of a difference quotient, by

$$\frac{\partial f(\underline{x})}{\partial x_i} \equiv \frac{\partial y}{\partial x_i} \equiv \lim_{\Delta x \to 0} \frac{f(x_1, \ldots x_i + \Delta x, \ldots x_{\hat{i}}) - f(\underline{x})}{\Delta x} \tag{20}$$

where the $\underline{x} - \{x_i\}$ are treated as constants, we don't have the constraint

$$(x_1, \ldots x_i + \Delta x, \ldots x_{\hat{i}}) \in \Im\langle f(\underline{x})\rangle \tag{21}$$

because the substitution of x_i with $x_i + \Delta x$ is a mere mathematical artifice finalized to definition (20), and whose second equivalence has sense only inasmuch is introduced the $y = f(\underline{x})$.

These definitions, in the case of $y = f(\underline{x})$ specified as $y = f(x)$ of which $x \equiv \{\underline{x} \parallel \hat{i} = 1\}$, are specified by

$$dy \equiv df(x) \equiv f'(x)dx \qquad dx \equiv \lim_{\Delta x \to 0} \Delta x$$

$$\frac{df(x)}{dx} \equiv f'(x) \equiv \frac{dy}{dx} \equiv \lim_{\Delta x \to 0} \frac{f(x + \Delta x) - f(x)}{\Delta x} \tag{22}$$

and are coherent with

$$\lim_{x \to x_0} x \equiv x_0 \pm dx.$$

A $\partial f_c(\underline{y})/\partial x_i$ is defined only if exists a $\underline{y}(\underline{x})$. Indeed, from: $\underline{y}(\underline{x})$; rule of derivation of a composite function; follows

$$\frac{\partial f_c(\underline{y})}{\partial x_i} = \frac{\partial f_c(\underline{y}(\underline{x}))}{\partial x_i} = \sum_{m=1}^{\hat{m}} \frac{\partial f_c(\underline{y})}{\partial y_m} \frac{\partial y_m(\underline{x})}{\partial x_i}$$

that for $\hat{i} = 1$ becomes

$$\frac{df_c(\underline{y})}{dx} = \frac{df_c(\underline{y}(x))}{dx} \equiv f_c'(\underline{y}(x)) = \sum_{m=1}^{\hat{m}} \frac{\partial f_c(\underline{y})}{\partial y_m} y_m'(x) \tag{23}$$

1.3.2 Discretization of an Euclidean space

Being $\underline{A} \times \underline{B}$ the Cartesian product of the two sets $\underline{A}$ and $\underline{B}$, we have

$$\prod_{k=1}^{\hat{k}} \underline{A}_k \equiv \underline{A}_1 \times \underline{A}_2 \times \ldots \underline{A}_{\hat{k}}$$

where the $\{\underline{A}_k; k = 1, \hat{k}\}$ are $\hat{k}$ sets and is applicable the associative property.

An $\hat{\mathrm{i}}$-dimensional hypercube is named $\hat{\mathrm{i}}$-cube and is point, rectilinear segment, square, cube, hypercube for $\hat{\mathrm{i}} = 0$, $\hat{\mathrm{i}} = 1$, $\hat{\mathrm{i}} = 2$, $\hat{\mathrm{i}} = 3$, $\hat{\mathrm{i}} > 3$.

Even if every point of $\mathcal{E}_{\hat{\mathrm{i}}}$ is of accumulation for each set whose boundary contains it, the adjacency between points of $\mathcal{E}_{\hat{\mathrm{i}}}$ cannot be contiguity i.e. contact, because, if so, the distance between two points would be null and consequently no part of $\mathcal{E}_{\hat{\mathrm{i}}}$ could have extent inasmuch a sum of null distances remain null even if the number of addends is infinite.

Therefore the adjacency between points of $\mathcal{E}_{\hat{\mathrm{i}}}$ is understood, not as contact, but as infinitesimal distance of type $\lim_{\Delta x \to 0} \Delta x$. From this follows the concrete discretization of such space, consisting in excluding from it each point that does not belong to the set of vertex of equal $\hat{\mathrm{i}}$-cubes each specification of the δV defined by

$$\delta V \equiv \lim_{\Delta x \to 0} \prod_{i=1}^{\hat{i}} [x_i, x_i + \Delta x] \equiv \prod_{i=1}^{\hat{i}} [x_i, x_i + dx_i] \tag{24}$$

whose measure $d\underline{x}$ is expressed by

$$d\underline{x} = dx_1 dx_2 \ldots dx_{\hat{i}} = \prod_{i=1}^{\hat{i}} dx_i$$

and so is infinitesimal of order $\hat{\mathrm{i}}$.

The boundary of δV is an its part constituted by $\hat{\mathrm{i}}$ couples of $(\hat{\mathrm{i}} - 1)$-cubes, as the $\{\delta A_i, \delta A_{+i}\}$ reciprocally distant dx_i in the i-th direction and whose δA_{+i} constitutes also the boundary and is part of the $(\hat{\mathrm{i}} - 1)$-cube infinitesimal δV_{+i} adjacent to δV in the sense of x_i increasing. On the boundary of δV there are $2^{\hat{i}}$ vertices with $\underline{x}$ the one that has minimum the value of each coordinate and lies also on δA_i, and $\hat{i}\, 2^{\hat{i}-1}$ rectilinear segments whose i-th conjoin two vertices in i-th direction. Indeed the $(i-1)$-cubes that lie on said boundary are $2^{\hat{i}-1}\binom{\hat{i}}{i}$, as we deduce by the being δV a specification of the parallelotope (generalization $\hat{\mathrm{i}}$-dimensional of a three-dimensional parallelepiped) in [91].

Therefore $\mathcal{E}_{\hat{\mathrm{i}}}$ has the discretization consisting in considering it divided in an infinite set of elements that specifying δV, as well as in agreeing, based on such condition, that a $f(\underline{x})$ has an only same value associated to δV homogeneously, i.e. equally in all its parts, and (excluding the case that is a constant function) has jumps (discontinuity of the first kind) infinitesimal on its boundary. This relation between $f(\underline{x})$ and δV can be only a mere association inasmuch f not describes further $\delta\mathrm{V}$ or it can also be that f describes only a part of the boundary of $\delta\mathrm{V}$. If f describes $\delta\mathrm{V}$ as an its property due to the particular nature of $\boldsymbol{\mathcal{E}}_{\hat{\mathrm{i}}}$, then it can be considered as average value of an infinite set of infinitesimals of second order homogeneously distributed in $\delta\mathrm{V}$ inasmuch biunivocally associated with just as many points of this. Analogous considerations are extended directly to portions $(\hat{\mathrm{i}}-\mathrm{n})$-dimensional of $\boldsymbol{\mathcal{E}}_{\hat{\mathrm{i}}}$.

From $Æ\langle \mathrm{d}\mathbf{x}, \mathrm{dx_i} \,/\!\!/\, \mathbf{a}, \mathbf{a}_\mathrm{i} \,/\!\!/\, (5), (7)\rangle$ follow

$$\mathrm{d}\mathbf{x} = \sum_{\mathrm{i}=1}^{\hat{\mathrm{i}}} \mathrm{dx_i}\hat{\boldsymbol{e}}_\mathrm{i} \qquad |\mathrm{d}\mathbf{x}| = \sqrt{\sum_{\mathrm{i}=1}^{\hat{\mathrm{i}}} \mathrm{dx_i^2}} \tag{25}$$

with $|\mathrm{d}\mathbf{x}|$ the length of the diagonal of $\delta\mathrm{V}$ i.e. distance between $\underline{\mathrm{x}}$ and the vertex that has maximum the value of every coordinate.

1.3.3 Integrals

Naming $\underline{\mathrm{e}}$ contingently present variables, the integral of function $g(\underline{\mathrm{x}}, \underline{\mathrm{e}})$, over integration domain V of which $\mathrm{V} \subseteq \boldsymbol{\mathcal{E}}_{\hat{\mathrm{i}}}$, is indicated $\int_\mathrm{V} g(\underline{\mathrm{x}}, \underline{\mathrm{e}})\, \mathrm{dV}$ of which

$$\mathrm{dV} \equiv \delta\mathrm{V} \equiv \mathrm{d}\underline{\mathrm{x}}$$

$$\int_\mathrm{V} g(\underline{\mathrm{x}}, \underline{\mathrm{e}})\, \mathrm{dV} \equiv \underset{\mathrm{V}}{\int\cdots\int} g(\underline{\mathrm{x}}, \underline{\mathrm{e}})\, \mathrm{dV} \equiv \int_\mathrm{V} g\, \mathrm{dV} \equiv \int_\mathrm{V} g(\underline{y}, \underline{\mathrm{e}})\, \mathrm{d}U \tag{26}$$

in whose third integral, the independent variables are absent because known, and whose $\underline{y}$ and U are two arbitrary names that do not need definitions because are sufficient $g(\underline{\mathrm{x}}, \underline{\mathrm{e}})$ and V for deducing $(\underline{y}, U) \equiv (\underline{\mathrm{x}}, \mathrm{V})$.

The definition of $\int_\mathrm{V} g(\underline{\mathrm{x}}, \underline{\mathrm{e}})\, \mathrm{dV}$ as limit of a sum as the number of its addends approaches infinity (i.e. a number unlimitedly large), allows us to introduce

$$\underset{\mathrm{V}}{\int\cdots\int} g(\underline{\mathrm{x}}, \underline{\mathrm{e}})\, \mathrm{dV} \equiv \sum_{\mathrm{c}=1}^{\infty} \delta\mathrm{M_c} \tag{27}$$

of which

$$\delta\mathrm{M_c} \equiv g(\underline{\mathrm{x}}_\mathrm{c}, \underline{\mathrm{e}})\delta\mathrm{V_c} \qquad \text{with} \qquad Æ\langle \underline{\mathrm{x}}_\mathrm{c}, \delta\mathrm{V_c} \,/\!\!/\, \underline{\mathrm{x}}, \delta\mathrm{V}\rangle,$$

$\{\delta\mathrm{V_c}; \mathrm{c} = 1, \infty\}$ a decomposition of V inasmuch verifies

$$\mathrm{V} = \bigcup_{\mathrm{c}=1}^{\infty} \delta\mathrm{V_c} \qquad \text{and} \qquad \{\delta\mathrm{V_a} \cap \delta\mathrm{V_b} = \varnothing; \forall \mathrm{a} \neq \mathrm{b}\},$$

and being so $\delta \mathrm{M_c}$ the product between the measure $\mathrm{d}\underline{\mathrm{x}}_\mathrm{c}$ of the infinitesimal portion $\delta \mathrm{V_c}$ of V and the value of $g(\underline{\mathrm{x}}, \underline{\mathrm{e}})$ in such portion.

The (27) highlights that a $\int_{\mathrm{V}-V} g\,\mathrm{d}\underline{\mathrm{x}}$, of which $V \subseteq \boldsymbol{\mathcal{E}}_{\hat{\mathrm{i}}}$, is equivalent to $\sum_{\mathrm{c}=1}^{\infty} \delta \mathrm{M_c}$ from which are eliminated the elements present also in the summation that (analogously to (27)) expresses $\int_V g\,\mathrm{d}\underline{\mathrm{x}}$. Therefore we have

$$\int \cdots \int_{\mathrm{V}-V} g\,\mathrm{d}\underline{\mathrm{x}} \equiv \sum_{\mathrm{c}=1}^{\infty} \delta M_\mathrm{c} \tag{28}$$

of which

$$\{\delta M_\mathrm{c}; \mathrm{c} = 1, \infty\} \equiv \underline{M} = \underline{\mathrm{M}} - \underline{\mathrm{N}}, \qquad \underline{\mathrm{M}} \equiv \{\delta \mathrm{M_c}; \mathrm{c} = 1, \infty\},$$

$$\underline{\mathrm{N}} \equiv \{\delta \mathrm{N_c}; \mathrm{c} = 1, \infty\}, \qquad \int_V g\,\mathrm{d}\underline{\mathrm{x}} \equiv \sum_{\mathrm{c}=1}^{\infty} \delta \mathrm{N_c}.$$

From: (27), last of previous; eliminating the couples of equal elements respectively present in the two summations, understand $\{\delta N_\mathrm{c}; \mathrm{c} = 1, \infty\} = \underline{\mathrm{N}} - \underline{\mathrm{M}}$; (28); follows

$$\begin{aligned}&\int_\mathrm{V} g\,\mathrm{d}\underline{\mathrm{x}} - \int_V g\,\mathrm{d}\underline{\mathrm{x}} \equiv \sum_{\mathrm{c}=1}^{\infty} \delta \mathrm{M_c} - \sum_{\mathrm{c}=1}^{\infty} \delta \mathrm{N_c} = \sum_{\mathrm{c}=1}^{\infty} \delta M_\mathrm{c} - \sum_{\mathrm{c}=1}^{\infty} \delta N_\mathrm{c} = \\ &\int_{\mathrm{V}-V} g\,\mathrm{d}\underline{\mathrm{x}} - \int_{V-\mathrm{V}} g\,\mathrm{d}\underline{\mathrm{x}}\end{aligned} \tag{29}$$

Chapter 2

Matter and physical space

We use $\mathfrak{m}$ as name of all matter. The temporal becoming is time pass. The material world is $\mathfrak{m}$ in temporal becoming; the volumic space $\mathcal{V}$ is the place of $\mathfrak{m}$; $\mathcal{V}$ in temporal becoming is the place of material world i.e. four-dimensional physical space $\mathcal{S}$ i.e. spacetime i.e. spatial-temporal continuum divisible in unlimitedly small parts.

Indeed we consider $\mathcal{V}$ and temporal becoming compliant to normal sensorial perception: $\mathcal{V}$ is represented as the particular $\mathcal{E}_{\hat{i}}$ whose $\underline{x}$ are specified by the volumic $\underline{x}$ of which

$$\underline{x} \equiv (x_i; i = 1, 3) \qquad \text{and} \qquad \mathbb{R}\langle x_i \rangle = \mathbb{R} \equiv (-\infty, \infty);$$

the temporal becoming of an object is represented as a succession of its positions on a straight line where a temporal abscissa is measured.

Thus representing $\mathcal{S}$ as such a succession of positions of $\mathcal{V}$, we have

$$\mathcal{S} = \mathcal{V} \times \mathbb{R}\langle t \rangle$$

with t the temporal coordinate of which $\mathbb{R}_t = \mathbb{R}$, and $\mathcal{S}$ is the $\mathcal{E}_{\hat{i}}$ whose $\underline{x}$ and $f(\underline{x})$ are specified by the $\underline{ж}$ of which

$$\underline{ж} \equiv (ж_i; i = 1, 4) \equiv (\underline{x}, t)$$

and by the $f(\underline{ж})$ of which

$$\mathfrak{I}\langle f(\underline{ж}) \rangle = \mathbb{R}\langle \underline{ж} \rangle = \mathbb{R}^4 \qquad \text{due to} \qquad \mathbb{R}\langle ж_i \rangle = \mathbb{R}.$$

The four values of $\underline{ж}$ are the default of the respective variables.

Moreover the tensors, introduced in section 1.2 as inherent the $\hat{i}$-dimensional $\mathcal{E}_{\hat{i}}$, are implicitly inherent the three-dimensional $\boldsymbol{\mathcal{V}}$, so a tensor of order $\mathbf{o}$ and a vector have respectively $3^{\mathbf{o}}$ and 3 components, and the $(\hat{\boldsymbol{e}}_i; i = 1, 3)$ of which $\hat{\boldsymbol{e}}_i \equiv \delta_{ij}$ are the coordinate versors of the reference system of $\boldsymbol{\mathcal{V}}$.

We name $\boldsymbol{\mathcal{V}}_t$ the place of $\boldsymbol{\mathfrak{m}}$ at the instant t i.e. the subset of $\boldsymbol{s}$ individuated by the instant t inasmuch constituted by the elements that have the same value t of the temporal coordinate.

The $\boldsymbol{s} = \boldsymbol{\mathcal{V}} \times \mathbb{R}_t$ shows $\boldsymbol{s}$ as the infinite set of each $\boldsymbol{\mathcal{V}}_t$ and $\boldsymbol{\mathcal{V}}$ as the generic element of such set. So $\boldsymbol{\mathcal{V}}$ is implicitly $\boldsymbol{\mathcal{V}}_t$ in coherence with being the value of t that of default.

A volume is a part of $\boldsymbol{\mathcal{V}}$ that has extent and shape defined by its boundary i.e. its external surface, that, inasmuch closed and without border, delimits it by separating it from its surroundings. The versor of a flat portion of a boundary is normal to it and is conventionally directed towards the outside of the volume. The measure of a volume $\mathbf{v}$ is $\int_v d\mathbf{v}$. A contact between two volumes takes place on a contact surface that is an intersection of the two sets constituted by the respective boundaries.

The (24) is specified by $\delta v \equiv \prod_{i=1}^{3}[x_i, x_i + dx_i]$ whereby δv is the immobile infinitesimal cube, identified by $\underline{x}$ as that of its eight vertices that has minimum the value of each coordinate, and whose boundary is the union $\underline{\delta A}$ of the six squares $\{\delta A_i, \delta A_{+i}; i = 1, 3\}$ with $\underline{x}$ vertex of δA_i, $-\hat{\boldsymbol{e}}_i$ and $\hat{\boldsymbol{e}}_i$ the respective versors of δA_i and δA_{+i} distant dx_i, and δA_{+i} that constitutes also the boundary of the infinitesimal cube δv_{+i} contiguous to δv in the sense of x_i increasing. So $\boldsymbol{\mathcal{V}}$ has the discretization constituted by its equivalence to the infinite set of infinitesimal cubes whose generic element is δv i.e. constituted of considering it divided in such elements.

In compliance with the know paradox whereby a set is infinite (and is also named "Dedekind-infinite set") if it is in biunivocal correspondence with an its proper subset consequently also this infinite, besides $\boldsymbol{\mathcal{V}}$ is an infinite set also a volume of finite extent. Moreover, in coherence with this and $\underline{P}_s \subset \underline{P}_v$ with $\underline{P}_v$ and $\underline{P}_s$ the sets of points of a volume and its boundary, also $\underline{P}_s$ is an infinite set as $\underline{P}_v$, but $\underline{P}_s$ has order lesser of a unity with respect that of $\underline{P}_v$ (being the order of infinity of a set that of its numerosity i.e. of the number of elements that constitute it).

A point $\underline{x}$ is inner to a domain (closed set) $\textsc{i}$ of which $\textsc{i} \subset \mathcal{E}_{\hat{i}}$, if $\exists D_{\underline{x}} \subset \textsc{i}$ with $D_{\underline{x}}$ a circular domain of center $\underline{x}$. Such an $\textsc{i}$ is internally connected if two arbitrarily chosen its inner points can be ever joint by a polygonal chain whose points are all inner to $\textsc{i}$.

A PdM (i.e. a part of $\mathfrak{m}$) has mass and volume in the sense that the first is its quantity and the second is the place of $\mathcal{V}$ where only it is collocated. Is implicit the equivalence between the names of a PdM, of its mass and of its volume.

2.1 Continuity in Continuum Mechanics

Also $\mathfrak{m}$ is considered a continuum as $\mathcal{S}$ (this gives the name to Continuum Mechanics). So inherently a PdM of mass M and volume V internally connected and devoid of holes, without temporal becoming i.e. in $\mathtt{t}$ we establish

$$\lim_{\mathsf{V}\to 0} \frac{\mathsf{M}}{\mathsf{V}} = \frac{\delta\mathrm{M}}{\delta\mathrm{V}} \equiv \rho(\underline{ж}) \neq \{0 \vee \infty\} \tag{30}$$

with ρ the mass density, C the PdM that has mass $\delta\mathrm{M}$ and volume $\delta\mathrm{V}$ of which $\lim_{\mathsf{V}\to 0} \mathsf{V} \equiv \delta\mathrm{V}$ and $\lim_{\mathsf{V}\to 0} \mathsf{M} \equiv \delta\mathrm{M}$.

In case $f(\underline{ж})$ expresses a property of $\mathfrak{m}$, is essentially (30) that consents the implicit

$$\left\{f(\underline{ж}) \in \mathsf{C}^0(\underline{ж}_0);\, \forall (f(\underline{ж}), \underline{ж}_0 \in \mathcal{S})\right\}.$$

The relation between $f(\underline{x})$ and $\delta\mathrm{V}$ in section 1.3.2 is specified by that between $f(\underline{ж})$ and $\delta\mathrm{V}$, being so f generally a physical quantity that describes C and is an its property.

In applications that represent cases of PdM, the limit of (30) is substituted by a V small enough to not be able to distinguish its parts with different values of ρ.

A body is the PdM contained in a volume whose boundary is not traversable by matter. Indeed a body in the passage of time has variable volume but constant mass coherently with the fundamental law of mass conservation (e.g. [24]).

2.2 Discretization of matter

An infinitesimal corpuscle is a body that has mass and volume both infinitesimal. The knowledge of infinitesimal and infinite entities is limited by their inexistence in the material world. However, having to minimize the imposition of conditions, also in observance of the economy of thought prescribed by the fundamental scientific rule know as *Novacula Occami*, let's admit that an infinitesimal corpuscle would appear to an imaginary infinitesimal observer (i.e. that would occupy an infinitesimal volume) in a completely equivalent way to how a finite body appear to a normal finite observer.

Coherently with this we attribute to $\boldsymbol{\mathfrak{m}}$ the discretization consistent in the consider it divided in (i.e. equivalent to) an infinite set of infinitesimal corpuscles, contiguous inasmuch every part of the boundary of one is ever in contact with one or more boundaries of others, generally different in mass and volume, and everyone mobile along a respective trajectory because, for the infinitesimal smallness of its volume, is indivisible and can't split up in mobile parts along respective trajectories distinguishable as different in terms of mutual finite distances. Therefore, analogously to a volume of finite extent, also the PdM contained in it is an infinite set of infinitesimal elements.

The generic element of such a set is the corpuscle $\mathfrak{c}$, mobile along the own trajectory $\mathcal{T}$, of constant mass $\delta\mathfrak{M}$ and variable volume $\delta\mathfrak{v}$.

Between the sets of each $\delta\mathrm{V}$ and $\delta\mathfrak{v}$, we introduce in $\mathtt{t}$ a biunivocal correspondence such that each couple has the shorter distance between the respective centroids.

Meaning that is $(\delta\mathrm{V}, \delta\mathfrak{v})$ the generic couple of such biunivocal correspondence, $\delta\mathrm{V}$ and $\delta\mathfrak{v}$ are generally different regarding position, extent and shape that could be seen by an imaginary infinitesimal observer. However these differences are unknowable to the human mind and consequently we can establish them at our discretion as long as they are compatible and agree with the given contextual conditions in which $\delta\mathrm{V}$ and $\delta\mathfrak{v}$ are infinitesimal. We imply and name $\mathcal{D}$ this subordinate discretion in establishing position, extension and form of $\delta\mathfrak{v}$ as well as of any other infinitesimal part of a space.

Therefore are implicit: $\delta\mathrm{V} \equiv \delta\mathfrak{v}$ and $\delta\mathrm{M} \equiv \delta\mathfrak{M}$ in $\mathtt{t}$, coherently with the be $\delta\mathrm{V}$ immobile and $\delta\mathfrak{v}$ mobile;

$$\delta\mathfrak{v} \equiv \prod_{i=1}^{3}[\mathfrak{x}_i, \mathfrak{x}_i + d\mathfrak{x}_i] \qquad \text{and} \qquad \underline{\mathfrak{x}} = \underline{\mathrm{x}}$$

with $\underline{\mathfrak{x}}$ (of which $\underline{\mathfrak{x}} \equiv (\mathfrak{x}_i; i = 1, 3)$) the vertex of $\delta\mathfrak{v}$ that has minimum the value of each coordinate; $\underline{\delta\mathfrak{A}}$ the boundary of $\delta\mathfrak{v}$ constituted by the union of the six infinitesimal squares

$$\{\delta\mathfrak{A}_i, \delta\mathfrak{A}_{+i}; i = 1, 3\},$$

being $\underline{\mathfrak{x}}$ vertex of $\delta\mathfrak{A}_i$ and with $-\hat{\boldsymbol{e}}_i$ and $\hat{\boldsymbol{e}}_i$ (of which $\hat{\boldsymbol{e}}_i \equiv \delta_{ij}$) the respective versors of $\delta\mathfrak{A}_i$ and $\delta\mathfrak{A}_{+i}$ distant $d\mathfrak{x}_i$.

On the basis of this, each immobile infinitesimal cube that specifies $\delta\mathrm{v}$ is in $\mathtt{t}$ (that is equivalent to $[\mathtt{t}, \mathtt{t} + \mathtt{dt}]$ in analogy to what has been said in section 1.3.2 about the relation between $f(\underline{\mathrm{x}})$ and $\delta\mathrm{v}$) the volume of a respective corpuscle mobile that specifies $\mathfrak{c}$.

The trajectory $\mathcal{T}$ of $\mathfrak{c}$ is the curve identified in a three-dimensional Euclidean space by the parametric equations $\underline{\mathfrak{x}} = \underline{\mathfrak{x}}(\mathtt{t})$ and to which corresponds the curve of $\mathcal{S}$ identified by the parametric equations

$$\{\underline{\mathfrak{x}} = \underline{\mathfrak{x}}(\mathtt{t}), \mathtt{t} = \mathfrak{x}_t(\mathtt{t}) \equiv \mathtt{t}\}.$$

Thus the position of $\mathfrak{c}$ in $\mathtt{t}$ can be named $\underline{\mathfrak{x}}$ or $\boldsymbol{\mathfrak{x}}_{\mathfrak{c}}$ of which $\boldsymbol{\mathfrak{x}}_{\mathfrak{c}} \equiv \mathfrak{x}_{\mathrm{i}}$ with such vector $\boldsymbol{\mathfrak{x}}_{\mathfrak{c}}$ applied to the origin of the three-dimensional reference system of $\boldsymbol{\mathcal{V}}$ and that has $\underline{\mathfrak{x}}$ as destination.

A tensor $\mathbf{A}$ in a function $h(\mathbf{A}, \underline{\mathrm{e}})$ is equivalent (in conformity to (1)) to the sequence of its components, hence $f_{\mathfrak{c}}(\underline{\mathfrak{x}}) \equiv f_{\mathfrak{c}}(\boldsymbol{\mathfrak{x}}_{\mathfrak{c}})$.

We introduce

$$f_{\mathfrak{c}}(\underline{\mathfrak{x}}), \qquad \mathfrak{I}\langle f_{\mathfrak{c}}(\underline{\mathfrak{x}})\rangle = \mathcal{T}, \qquad \underline{\mathfrak{x}}, \qquad \delta\mathfrak{v}$$

as specifications of the

$$f(\underline{\mathrm{x}}), \qquad \mathfrak{I}\langle f(\underline{\mathrm{x}})\rangle = \mathcal{E}_{\hat{\mathrm{i}}}, \qquad \underline{\mathrm{x}}, \qquad \delta\mathrm{v}$$

of section 1.3.2.

Based on this, the relation between $f(\underline{\mathrm{x}})$ and $\delta\mathrm{v}$ in such section is specified by that between $f_{\mathfrak{c}}(\underline{\mathfrak{x}})$ and $\delta\mathfrak{v}$, so we establish $f_{\mathfrak{c}}$ as physical quantity that describes $\mathfrak{c}$ and is an its property.

A $f_{\mathfrak{c}}(\underline{\mathfrak{x}})$ can only have values univocally corresponding to the positions that $\delta\mathfrak{v}$ occupies on $\mathcal{T}$ as $\mathtt{t}$ varies. Therefore these values are those that $\delta\mathfrak{v}$ transports with it during its motion and we mean this when we say that a quantity $f_{\mathfrak{c}}$ is transported by $\mathfrak{c}$: an imaginary infinitesimal observer, constantly positioned in $\mathfrak{c}$, at the instant $\mathtt{t}$ would measure $f_{\mathfrak{c}}$ as value (average of infinitesimals of second order) of the quantity f in such corpuscle. Is implicit that $f_{\mathfrak{c}}$ implies $f_{\mathfrak{c}}(\underline{\mathfrak{x}})$.

This, $\underline{\mathfrak{x}} = \underline{\mathrm{x}}$ at $\mathtt{t}$ and the being $\underline{\mathrm{x}}$ the coordinates of $\boldsymbol{\mathcal{V}}_{\mathtt{t}}$ allow us to introduce

$$\{f_{\mathfrak{c}}(\underline{\mathfrak{x}}) = f(\underline{ж}); \forall f_{\mathfrak{c}}\} \tag{31}$$

whose equation implies $f_{\mathfrak{c}} = f$, and of which $Æ\langle f_{\mathfrak{c}}(\underline{\mathfrak{x}}), f(\underline{ж}) \,/\!/\, f_{\mathrm{A}}(\underline{\mathrm{a}}), f_{\mathrm{B}}(\underline{\mathrm{b}}) \,/\!/ (18)\rangle$.

Regarding (31) is implicit that $f_{\mathfrak{c}}$ and f of $f(\underline{\text{ж}})$ can have arbitrary subscripts, except for the condition that the first name differs from the second only for the presence of $\mathfrak{c}$.

A volume is a thermodynamic system if is in a state of thermodynamic equilibrium i.e. if every thermodynamic quantity (e.g. the temperature) it is homogeneously distributed there i.e. it has a same value in every its point. A thermodynamic system is open or closed or isolated if can exchange with its surroundings energy and matter or only energy or neither energy nor matter.

An equation of state links thermodynamic quantities, is specifically own of a certain substance, is determined by means of experiments and/or submacroscopic models i.e. at the atomic-molecular scale. Two equations of state more known ([43]) are those thermal (that links density, temperature and thermodynamic pressure, and of which is often, as in the following, omitted such name) and caloric (that links the internal energy to two of the aforementioned).

Coherently with the relation between $f(\underline{\text{ж}})$ and $\delta \mathrm{v}$, the $\delta \mathrm{v}$ and $\delta \mathfrak{v}$ are thermodynamic systems, open $\delta \mathrm{v}$ and closed $\delta \mathfrak{v}$, because the physical quantities that describe them have in they (excluding infinitesimals of higher order) a same constant value in all their infinitesimal extent. This, which is referred as local thermodynamic equilibrium, implies that such quantities may appear in the relations own of Thermodynamics e.g. an equation of state.

Such $\delta \mathrm{v}$ and $\delta \mathfrak{v}$, as well as $f(\underline{\text{ж}})$ and $f_{\mathfrak{c}}(\underline{\mathbf{x}})$ of (31), correspond to the respective descriptions Eulerian (or spatial or local) and Lagrangian (or substantial or material) of a motion: in the first, specifications of $\delta \mathrm{v}$ are considered and the quantities are functions of $\underline{\text{ж}}$; in the second, specifications of $\delta \mathfrak{v}$ are considered and the quantities are functions of t i.e. of the parametric functions (with t independent variable) of the trajectories of such specifications.

2.3 Total, local (Eulerian) and material (Lagrangian) derivatives.

From: (20); (31), absence of the constraint that specifies (21), (18); follows

$$\begin{gathered}\frac{\partial f_{\mathfrak{c}}(\underline{\mathbf{x}})}{\partial \mathbf{x}_{\mathrm{i}}} \equiv \lim_{\Delta \mathbf{x} \to 0} \frac{f_{\mathfrak{c}}(\mathbf{x}_{\mathrm{j}} + \delta_{\mathrm{ij}} \Delta \mathbf{x}; \mathrm{j} = 1,3) - f_{\mathfrak{c}}(\underline{\mathbf{x}})}{\Delta \mathbf{x}} \equiv \\ \lim_{\Delta \mathrm{x} \to 0} \frac{f\big((\mathrm{x}_{\mathrm{j}} + \delta_{\mathrm{ij}} \Delta \mathrm{x}; \mathrm{j} = 1,3), \mathrm{t}\big) - f(\underline{\text{ж}})}{\Delta \mathrm{x}} = \frac{\partial f(\underline{\text{ж}})}{\partial \mathrm{x}_{\mathrm{i}}}\end{gathered} \tag{32}$$

From:

$$\Delta \boldsymbol{x}_c \equiv \boldsymbol{x}_c(\mathtt{t} + \Delta\mathtt{t}) - \boldsymbol{x}_c(\mathtt{t}); \qquad (22); \qquad þ; \qquad Æ\langle \mathbf{w} \,/\!\!/\, f \,/\!\!/\, (31)\rangle;$$

follows

$$\lim_{\Delta t \to 0} \frac{\Delta \boldsymbol{x}_c}{\Delta t} \equiv \lim_{\Delta t \to 0} \frac{\boldsymbol{x}_c(t + \Delta t) - \boldsymbol{x}_c(t)}{\Delta t} = \frac{d\boldsymbol{x}_c}{dt} \equiv \mathbf{w}_c(\underline{x}) = \mathbf{w}(\underline{ж}) \qquad (33)$$

with $\Delta\boldsymbol{x}_c$ the displacement of $\mathfrak{c}$ between the instants $\mathtt{t}$ and $\mathtt{t}+\Delta\mathtt{t}$, $d\boldsymbol{x}_c$ and $\mathbf{w}_c$ the infinitesimal displacement and the velocity of $\mathfrak{c}$, being these three vectors applied to the destination of $\boldsymbol{x}_c$. Relating to the vector $\mathbf{w}$, we introduce $\mathbf{w} \equiv \mathrm{w}_i$.

From: (23);

$$x_i'(t) = w_i(\underline{ж}) \text{ (due to } d\boldsymbol{x}_c/dt = \mathbf{w}(\underline{ж}) \text{ in (33)), (32); (14);}$$

follows

$$\frac{df_c(\underline{x})}{dt} = \sum_{i=1}^{3} \frac{\partial f_c(\underline{x})}{\partial x_i} x_i'(t) = \sum_{i=1}^{3} \frac{\partial f(\underline{ж})}{\partial x_i} w_i(\underline{ж}) = \nabla f(\underline{ж}) \cdot \mathbf{w}(\underline{ж}) \qquad (34)$$

From:

$$(23), \quad x_i'(t) = w_i(\underline{ж}); \qquad (14); \qquad (34);$$

follows

$$\frac{df(\underline{x},t)}{dt} = \frac{\partial f(\underline{ж})}{\partial t} + \sum_{i=1}^{3} \frac{\partial f(\underline{ж})}{\partial x_i} w_i(\underline{ж}) = \frac{\partial f(\underline{ж})}{\partial t} + \nabla f(\underline{ж}) \cdot \mathbf{w}(\underline{ж}) = \frac{\partial f(\underline{ж})}{\partial t} + \frac{df_c(\underline{x})}{dt} \qquad (35)$$

whose $f'(\underline{x}(t),t)$ is named total because sum of the other two derivatives, $\partial f(\underline{ж})/\partial t$ and $f_c'(\underline{x}(t))$, that have relevance to the point $\underline{ж}$ of $\mathcal{S}$ and that are respectively local (or Eulerian) and material (or Lagrangian) as the respective functions to be derived.

A derivative is also named variation, by meaning unitary variation i.e. per unit of variable of derivation. The $\partial f(\underline{ж})/\partial t$ and $f_c'(\underline{x}(t))$ are variations that occur in the respective $\delta\mathrm{v}$ and $\delta\mathfrak{v}$ (this difference between the two variations would be seen by the said infinitesimal observer). The sum of some variations, that occur in respective volumes, occurs necessarily in the union of such volumes, and so (35) shows that $f'(\underline{\mathbf{x}}(\mathtt{t}),\mathtt{t})$ is a variation that occurs in $\delta\mathrm{v} \cup \delta\mathfrak{v}$. Moreover $f'(\underline{\mathbf{x}}(\mathtt{t}),\mathtt{t})$ is the whole variation of f in $\delta\mathrm{v} \cup \delta\mathfrak{v}$ because in no part of this occur other variations besides the said two. This occurrence in $\delta\mathrm{v} \cup \delta\mathfrak{v}$ and entirety imply that $f(\underline{\mathbf{x}},\mathtt{t})$ expresses the f in $\delta\mathrm{v} \cup \delta\mathfrak{v}$.

These distinctions, as well as that between $\delta\mathfrak{v}$ and $\delta\mathrm{v}$ (of which $\delta\mathfrak{v} \equiv \delta\mathrm{v}$ in $\mathtt{t}$), they would be impossible if they concerned only the instant $\mathtt{t}$, but their validity is instead highlighted by having to consider, imposed by (17) and implicit continuity of every functions, not the point $\mathtt{t}$ but limits approaching it.

Chapter 3

Transport theorem and balance equation

The generic infinitesimal portion $\delta\mathsf{S}$ of the boundary S of the volume V has the versor $\hat{\mathbf{e}}_{\mathsf{S}}$, of which $\hat{\mathbf{e}}_{\mathsf{S}} \equiv \hat{\mathrm{e}}_{\mathsf{S}\mathrm{i}}$, that is normal to $\delta\mathsf{S}$ and that, as for the boundary of each volume, is external i.e. oriented towards to the outside of V.

In coherence with (26) and $\underline{\mathsf{P}}_{\mathsf{S}} \subset \underline{\mathsf{P}}_{\mathsf{V}}$ of chapter 2, is implicit that the integrand function of an integral over a volume or surface has the $\underline{ж}$ as independent variables.

The divergence (or Gauss's or Ostrogradsky's) theorem is expressed by

$$\iiint_{\mathsf{V}} \nabla \cdot \mathbf{A} \, \mathrm{d}\mathsf{V} = \iint_{\mathsf{S}} \hat{\mathbf{e}}_{\mathsf{S}} \cdot \mathbf{A} \, \mathrm{d}\mathsf{S} \tag{36}$$

i.e.

$$\iiint_{\mathsf{V}} \sum_{\mathrm{i}=1}^{3} \frac{\partial A_{\mathrm{i}\mathrm{i}_2 \ldots \mathrm{i}_o}}{\partial \mathrm{x}_\mathrm{i}} \, \mathrm{d}\mathsf{V} = \iint_{\mathsf{S}} \sum_{\mathrm{i}=1}^{3} \hat{\mathrm{e}}_{\mathsf{S}\mathrm{i}} A_{\mathrm{i}\mathrm{i}_2 \ldots \mathrm{i}_o} \, \mathrm{d}\mathsf{S}$$

i.e.

$$\iiint_{\mathsf{V}} \frac{\partial \mathrm{F}}{\partial \mathrm{x}_\mathrm{i}} \, \mathrm{d}\mathsf{V} = \iint_{\mathsf{S}} \hat{\mathrm{e}}_{\mathsf{S}\mathrm{i}} \mathrm{F} \, \mathrm{d}\mathsf{S}$$

3.1 Reynolds transport theorem

From (20) and the fact that the integral of a limit is equal to the limit of the integral if such limit influences only the integrand function, follows

$$\lim_{\Delta \mathrm{t} \to 0} \iiint_{\mathsf{V}} \frac{f(\underline{ж})}{\Delta \mathrm{t}} \, \mathrm{d}\mathsf{V} = \lim_{\Delta \mathrm{t} \to 0} \iiint_{\mathsf{V}} \left(\frac{f(\underline{\mathrm{x}}, \mathrm{t} + \Delta \mathrm{t})}{\Delta \mathrm{t}} - \frac{\partial f(\underline{ж})}{\partial \mathrm{t}} \right) \mathrm{d}\mathsf{V} \tag{37}$$

We name $\mathfrak{C}$ a body of mass $\mathfrak{m}$ and volume $\mathfrak{v}$ devoid of holes, internally connected and that has boundary $\mathfrak{s}$. Such $\mathfrak{v}$ and $\mathfrak{s}$ at the instant t are indicated $\mathfrak{v}_t$ and $\mathfrak{s}_t$ of which $\mathfrak{v}_t \subset \boldsymbol{\mathcal{V}}_t$. This and the being implicit the value of t make implicit also $\mathfrak{v} \equiv \mathfrak{v}_\mathrm{t}$ and $\mathfrak{s} \equiv \mathfrak{s}_\mathrm{t}$.

From: (22), substitution of $\mathfrak{v}_{\mathrm{t}+\Delta\mathrm{t}}$ with the volume $\mathcal{v}_{\mathrm{t}+\Delta\mathrm{t}}$ that differs from $\mathfrak{v}_{\mathrm{t}+\Delta\mathrm{t}}$ only in consequence of $\mathcal{v}_{\mathrm{t}+\Delta\mathrm{t}} \subset \boldsymbol{\mathcal{V}}_\mathrm{t}$ and $\mathfrak{v}_{\mathrm{t}+\Delta\mathrm{t}} \subset \boldsymbol{\mathcal{V}}_{\mathrm{t}+\Delta\mathrm{t}}$; (37); follows

$$
\begin{aligned}
&\frac{\mathrm{d} \iiint_{\mathfrak{v}} f(\underline{\text{ж}})\, \mathrm{d}\mathfrak{v}}{\mathrm{dt}} = \\
&\lim_{\Delta\mathrm{t}\to 0} \frac{\iiint_{\mathcal{v}_{\mathrm{t}+\Delta\mathrm{t}}} f(\underline{\mathrm{x}}, \mathrm{t}+\Delta\mathrm{t})\, \mathrm{d}\mathfrak{v} - \iiint_{\mathfrak{v}_\mathrm{t}} f(\underline{\text{ж}})\, \mathrm{d}\mathfrak{v}}{\Delta\mathrm{t}} = \\
&\iiint_{\mathfrak{v}_\mathrm{t}} \frac{\partial f(\underline{\text{ж}})}{\partial \mathrm{t}}\, \mathrm{d}\mathfrak{v} + \lim_{\Delta\mathrm{t}\to 0} \frac{\iiint_{\mathcal{v}_{\mathrm{t}+\Delta\mathrm{t}}} f(\underline{\mathrm{x}}, \mathrm{t}+\Delta\mathrm{t})\, \mathrm{d}\mathfrak{v} - \iiint_{\mathfrak{v}_\mathrm{t}} f(\underline{\mathrm{x}}, \mathrm{t}+\Delta\mathrm{t})\, \mathrm{d}\mathfrak{v}}{\Delta\mathrm{t}}
\end{aligned}
\tag{38}
$$

The generic infinitesimal portion of $\mathfrak{s}$ is the areola $\delta\mathfrak{s}$ that has versor $\hat{\mathbf{e}}_\mathfrak{s}$, position vector $\boldsymbol{\mathcal{X}}_\mathfrak{s}$ applied to the origin of the reference system of $\boldsymbol{\mathcal{V}}$, and displacement vector $\Delta\boldsymbol{\mathcal{X}}_\mathfrak{s}$ applied to the destination of $\boldsymbol{\mathcal{X}}_\mathfrak{s}$ and accomplished between the instants t and $\mathrm{t}+\Delta\mathrm{t}$.

By naming $\mathbf{h}\delta\mathfrak{s}$ and $\mathbf{k}\delta\mathfrak{s}$ the generic elements of the decompositions

$$
\{\mathbf{h}_\mathrm{c}\delta\mathfrak{s}_\mathrm{c}; \mathrm{c}=1,\infty\} \qquad \text{and} \qquad \{\mathbf{k}_\mathrm{c}\delta\mathfrak{s}_\mathrm{c}; \mathrm{c}=1,\infty\}
$$

of $\mathcal{v}_{\mathrm{t}+\Delta\mathrm{t}} - \mathfrak{v}_\mathrm{t}$ and $\mathfrak{v}_\mathrm{t} - \mathcal{v}_{\mathrm{t}+\Delta\mathrm{t}}$, in the sense of

$$
\mathcal{v}_{\mathrm{t}+\Delta\mathrm{t}} - \mathfrak{v}_\mathrm{t} = \bigcup_{\mathrm{c}=1}^{\infty} \mathbf{h}_\mathrm{c}\delta\mathfrak{s}_\mathrm{c} \qquad \{\mathbf{h}_\mathrm{a}\delta\mathfrak{s}_\mathrm{a} \cap \mathbf{h}_\mathrm{b}\delta\mathfrak{s}_\mathrm{b} = \varnothing; \forall \mathrm{a} \neq \mathrm{b}\}
$$

and of

$$
\mathfrak{v}_\mathrm{t} - \mathcal{v}_{\mathrm{t}+\Delta\mathrm{t}} = \bigcup_{\mathrm{c}=1}^{\infty} \mathbf{k}_\mathrm{c}\delta\mathfrak{s}_\mathrm{c} \qquad \{\mathbf{k}_\mathrm{a}\delta\mathfrak{s}_\mathrm{a} \cap \mathbf{k}_\mathrm{b}\delta\mathfrak{s}_\mathrm{b} = \varnothing; \forall \mathrm{a} \neq \mathrm{b}\},
$$

we have

$$
\lim_{\Delta\mathrm{t}\to 0} \mathbf{h} \equiv \lim_{\Delta\mathrm{t}\to 0} \mathbf{k} \equiv \lim_{\Delta\mathrm{t}\to 0} |\mathbf{g}|
\tag{39}
$$

of which $\mathbf{g} \equiv \Delta\boldsymbol{\mathcal{X}}_\mathfrak{s} \cdot \hat{\mathbf{e}}_\mathfrak{s}$, as well as

$$
\begin{gathered}
\lim_{\Delta\mathrm{t}\to 0} ((\mathbf{h}\delta\mathfrak{s}, \mathcal{v}_{\mathrm{t}+\Delta\mathrm{t}} - \mathfrak{v}_\mathrm{t}), (\mathbf{k}\delta\mathfrak{s}, \mathfrak{v}_\mathrm{t} - \mathcal{v}_{\mathrm{t}+\Delta\mathrm{t}})) \Rightarrow \\
\text{Æ}\langle(\mathbf{h}\delta\mathfrak{s}, \mathcal{v}_{\mathrm{t}+\Delta\mathrm{t}} - \mathfrak{v}_\mathrm{t}), (\mathbf{k}\delta\mathfrak{s}, \mathfrak{v}_\mathrm{t} - \mathcal{v}_{\mathrm{t}+\Delta\mathrm{t}}) \,/\!/\, \delta\mathrm{v}, \mathrm{v} \,/\!/\, (27)\rangle
\end{gathered}
$$

that entails

$$\lim_{\Delta t \to 0} \iiint_{\mathfrak{v}_{t+\Delta t}-\mathfrak{v}_t} \mathrm{y}\, d\mathfrak{v} = \lim_{\Delta t \to 0} \iint_{\mathfrak{s}^+} \mathrm{y}\mathbf{h}\, d\mathfrak{s}$$
$$\lim_{\Delta t \to 0} \iiint_{\mathfrak{v}_t-\mathfrak{v}_{t+\Delta t}} \mathrm{y}\, d\mathfrak{v} = \lim_{\Delta t \to 0} \iint_{\mathfrak{s}^-} \mathrm{y}\mathbf{k}\, d\mathfrak{s} \tag{40}$$

of which $\mathrm{y}(\underline{\mathrm{x}}, \underline{\mathrm{e}})$, and whose $\mathfrak{s}^+$ and $\mathfrak{s}^-$ are the respective parts of $\mathfrak{s}$ where $\varpi_{\mathbf{g}} = 1$ and $\varpi_{\mathbf{g}} = -1$ being Æ$\langle \mathbf{g} \,/\!/\, \mathfrak{s} \,/\!/\, (8) \rangle$.

From: (29); (40); (39); the being $\mathfrak{s}^+ \cup \mathfrak{s}^-$ the part of $\mathfrak{s}$ where $\mathbf{g} \neq 0$, $\mathfrak{s}^+ \cap \mathfrak{s}^- = \varnothing$, additive property of the integral; follows

$$\lim_{\Delta t \to 0} \left(\iiint_{\mathfrak{v}_{t+\Delta t}} \mathrm{y}\, d\mathfrak{v} - \iiint_{\mathfrak{v}_t} \mathrm{y}\, d\mathfrak{v} \right) = \lim_{\Delta t \to 0} \left(\iiint_{\mathfrak{v}_{t+\Delta t}-\mathfrak{v}_t} \mathrm{y}\, d\mathfrak{v} - \iiint_{\mathfrak{v}_t-\mathfrak{v}_{t+\Delta t}} \mathrm{y}\, d\mathfrak{v} \right) =$$
$$\lim_{\Delta t \to 0} \left(\iint_{\mathfrak{s}^+} \mathrm{y}\mathbf{h}\, d\mathfrak{s} - \iint_{\mathfrak{s}^-} \mathrm{y}\mathbf{k}\, d\mathfrak{s} \right) = \lim_{\Delta t \to 0} \left(\iint_{\mathfrak{s}^+} \mathrm{y}\mathbf{g}\, d\mathfrak{s} + \iint_{\mathfrak{s}^-} \mathrm{y}\mathbf{g}\, d\mathfrak{s} \right) = \tag{41}$$
$$\lim_{\Delta t \to 0} \iint_{\mathfrak{s}} \mathrm{y}\mathbf{g}\, d\mathfrak{s}$$

From:

$$(41), \qquad \mathbf{g} \equiv \Delta \boldsymbol{x}_{\mathfrak{s}} \cdot \hat{\mathbf{e}}_{\mathfrak{s}}; \qquad \lim_{\Delta t \to 0} \frac{\Delta \boldsymbol{x}_{\mathfrak{s}}}{\Delta t} = \mathbf{W}(\underline{\text{ж}});$$

commutativity of the scalar product of the vectors $f\mathbf{W}$ and $\hat{\mathbf{e}}_{\mathfrak{s}}$;

$$Æ\langle \mathfrak{v}, \mathfrak{s}, f\mathbf{W} \,/\!/\, \mathrm{V}, \mathrm{S}, \mathbf{A} \,/\!/\, (36) \rangle;$$

follows

$$\lim_{\Delta t \to 0} \frac{\iiint_{\mathfrak{v}_{t+\Delta t}} f(\underline{\mathrm{x}}, t + \Delta t)\, d\mathfrak{v} - \iiint_{\mathfrak{v}_t} f(\underline{\mathrm{x}}, t + \Delta t)\, d\mathfrak{v}}{\Delta t} =$$
$$\lim_{\Delta t \to 0} \iint_{\mathfrak{s}} f(\underline{\mathrm{x}}, t + \Delta t) \frac{\Delta \boldsymbol{x}_{\mathfrak{s}}}{\Delta t} \cdot \hat{\mathbf{e}}_{\mathfrak{s}}\, d\mathfrak{s} =$$
$$\iint_{\mathfrak{s}} f\mathbf{W} \cdot \hat{\mathbf{e}}_{\mathfrak{s}}\, d\mathfrak{s} = \iint_{\mathfrak{s}} \hat{\mathbf{e}}_{\mathfrak{s}} \cdot f\mathbf{W}\, d\mathfrak{s} = \iiint_{\mathfrak{v}} \nabla \cdot f\mathbf{W}\, d\mathfrak{v}$$

This and (38) give rise to

$$\frac{\mathrm{d}\iiint_{\mathfrak{v}} f\,\mathrm{d}\mathfrak{v}}{\mathrm{d}t} = \iiint_{\mathfrak{v}} \frac{\partial f}{\partial t}\mathrm{d}\mathfrak{v} + \iint_{\mathfrak{s}} f\mathbf{w}\cdot\hat{\mathbf{e}}_{\mathfrak{s}}\,\mathrm{d}\mathfrak{s} = \iiint_{\mathfrak{v}}\left(\frac{\partial f}{\partial t} + \nabla\cdot f\mathbf{w}\right)\mathrm{d}\mathfrak{v} \qquad (42)$$

that expresses the Reynolds transport theorem.

3.2 Balance and conservation equations

The obvious balance principle (attributed in [4] to the same Osborne Reynolds of the previous theorem) affirms that the variation of a quantity F contained in the control volume V is equal to the sum of the F generated (created, produced) in V plus the F that enters in V trough its boundary S.

By applying such a principle to the F that has density per unit of volume f, we obtain the generic balance equation

$$\frac{\mathrm{d}\iiint_{V} f\,\mathrm{d}V}{\mathrm{d}t} = \iiint_{V} G_F\,\mathrm{d}V - \iint_{S} f\boldsymbol{w}_S\cdot\hat{\mathbf{e}}_S\,\mathrm{d}S - \iint_{S} \boldsymbol{\phi}_F\cdot\hat{\mathbf{e}}_S\,\mathrm{d}S$$

where G_F is the F generated per unit of volume and time, $\boldsymbol{w}_S = \mathbf{w} - \mathbf{w}_S$, $\mathbf{w}_S$ is the velocity of δS, $f\boldsymbol{w}_S$ is the flux (quantity per unit of time and surface) of F convective (caused by motion of matter i.e. by the F transported by the matter that cross δS) in the oriented direction of the relative velocity $\boldsymbol{w}_S$, $\boldsymbol{\phi}_F$ is the flux of F diffusive (independent by the motion of matter e.g. heat by thermal conduction), the sign "–" that precedes the fluxes is due to the conventions of $\hat{\mathbf{e}}_S$ directed towards the outside of V and to the consider positive a quantity acquired by a volume. Such an equation is a conservation law if

$$\frac{\mathrm{d}\iiint_{V} f\,\mathrm{d}V}{\mathrm{d}t} = 0$$

Chapter 4

Principle of conservation of mass

The discretization of $\mathfrak{m}$ (said in section 2.2) implies, for any PdM C that has mass M and volume V, the $\mathsf{C} \equiv \{\mathtt{c}_\mathrm{c}; \mathrm{c} = 1, \infty\}$ of which Æ$\langle \mathtt{c}_\mathrm{c} \,/\!/\, \mathtt{c} \rangle$ with $\mathtt{c}$ the infinitesimal corpuscle that has mass $\delta\mathrm{M}$ and volume $\delta\mathrm{V}$. Therefore the properties of $\mathtt{c}_\mathrm{c}$ are referred by adding the subscript "c" to the name of those of $\mathtt{c}$ and are specifications of these, e.g. $\delta\mathrm{M}_\mathrm{c}$ and $\delta\mathrm{V}_\mathrm{c}$ are mass and volume of $\mathtt{c}_\mathrm{c}$ of which Æ$\langle \delta\mathrm{M}_\mathrm{c}, \delta\mathrm{V}_\mathrm{c} \,/\!/\, \delta\mathrm{M}, \delta\mathrm{V} \rangle$. Based on this we have

$$\mathsf{V} \equiv \{\delta\mathrm{V}_\mathrm{c}; \mathrm{c} = 1, \infty\} \qquad \mathsf{V} = \bigcup_{\mathrm{c}=1}^{\infty} \delta\mathrm{V}_\mathrm{c} \qquad \mathsf{M} = \sum_{\mathrm{c}=1}^{\infty} \delta\mathrm{M}_\mathrm{c} \tag{43}$$

From: this; $\delta\mathrm{M} = \rho\delta\mathrm{V}$ (from (30));

$$\text{Æ}\langle \rho, \underline{\mathtt{x}}, \mathtt{t}, \delta\mathrm{V}, \mathsf{V} \,/\!/\, g, \underline{\mathtt{x}}, \underline{\mathtt{e}}, \delta\mathrm{V}, \mathrm{V} \,/\!/\, (27) \rangle;$$

follows

$$\mathsf{M} = \sum_{\mathrm{c}=1}^{\infty} \delta\mathrm{M}_\mathrm{c} = \sum_{\mathrm{c}=1}^{\infty} \rho(\underline{\mathtt{x}}_\mathrm{c}, \mathtt{t})\delta\mathrm{V}_\mathrm{c} = \iiint_{\mathsf{V}} \rho \, \mathrm{d}\mathsf{V} \tag{44}$$

From:

$$\text{Æ}\langle \mathfrak{m}, \mathfrak{v}, \mathfrak{C} \,/\!/\, \mathsf{M}, \mathsf{V}, \mathsf{C} \,/\!/\, (44) \rangle; \qquad \text{Æ}\langle \rho \,/\!/\, f \,/\!/\, (42) \rangle;$$

Æ$\langle \rho, \mathbf{w} \,/\!/\, A, \mathbf{B} \,/\!/\, (16) \rangle$, commutativity of the scalar product; follows

$$\frac{\mathrm{d}\mathfrak{m}}{\mathrm{d}\mathtt{t}} = \frac{\mathrm{d} \iiint_{\mathsf{V}} \rho \, \mathrm{d}\mathfrak{v}}{\mathrm{d}\mathtt{t}} = \iiint_{\mathsf{V}} \left(\frac{\partial \rho}{\partial \mathtt{t}} + \nabla \cdot \rho\mathbf{w} \right) \mathrm{d}\mathfrak{v} =$$

$$\iiint_{\mathsf{V}} \left(\frac{\partial \rho}{\partial \mathtt{t}} + \rho\nabla \cdot \mathbf{w} + \nabla\rho \cdot \mathbf{w} \right) \mathrm{d}\mathfrak{v}$$

This, $d\mathfrak{m}/dt = 0$ (because $\mathfrak{m}$ is the mass of a body), the genericity of $\mathfrak{v}$ (that consents to deduce equality of integrand functions from equality of integrals) and Æ⟨ρ // f // (35)⟩ give rise to

$$\frac{\partial \rho}{\partial t} + \nabla \cdot \rho \mathbf{w} = 0 \qquad \frac{d\rho}{dt} + \rho \nabla \cdot \mathbf{w} = 0 \tag{45}$$

that are two forms of the equation of continuity for the mass namely of the principle of conservation of mass.

Chapter 5

Deformation

Naming deformation of a volume the relative movements between its points, its result are the differences of extent and shape of such volume that it has caused, its velocity is the rapidity with which it takes place over time.

We consider the deformation of $\delta\mathfrak{v}$ between $\mathtt{t}-\Delta\mathtt{t}$ and $\mathtt{t}$ in order to describe it, on the basis of

$$\text{Æ}\langle f_{\mathfrak{c}}(\underline{\mathtt{x}}), \delta\mathfrak{v} \,/\!/\, f(\underline{\mathtt{x}}), \delta\mathtt{V} \,/\!/\, \text{section 1.3.2}\rangle$$

and (31), with functions of type $f(\underline{ж})$.

In coherence with Æ$\langle \boldsymbol{\mathfrak{x}} \,/\!/\, f \,/\!/\, (31)\rangle$, $\boldsymbol{\mathfrak{x}}_{\mathfrak{c}} \equiv \mathfrak{x}_i$ and (in $\mathtt{t}$) $\mathfrak{x}_i = \mathtt{x}_i$, we introduce $\boldsymbol{\mathfrak{x}} = \boldsymbol{\mathfrak{a}} + \boldsymbol{\mathfrak{s}}$ of which

$$\boldsymbol{\mathfrak{x}} \equiv \mathtt{x}_i, \quad \boldsymbol{\mathfrak{a}} \equiv \mathtt{a}_i, \quad \boldsymbol{\mathfrak{s}} \equiv \mathtt{s}_i, \quad \boldsymbol{\mathfrak{s}} = \boldsymbol{\mathfrak{s}}(\boldsymbol{\mathfrak{a}}) \vee \boldsymbol{\mathfrak{s}}(\boldsymbol{\mathfrak{x}}), \quad \boldsymbol{\mathfrak{x}} = x(\boldsymbol{\mathfrak{a}}), \quad \boldsymbol{\mathfrak{a}} = \boldsymbol{\mathfrak{a}}(\boldsymbol{\mathfrak{x}})$$

with $\boldsymbol{\mathfrak{a}}$ and $\boldsymbol{\mathfrak{x}}$ the positions of $\delta\mathfrak{v}$ in $\mathtt{t}-\Delta\mathtt{t}$ and $\mathtt{t}$, being so $\boldsymbol{\mathfrak{s}}$ the displacement of $\delta\mathfrak{v}$ in the interval $[\mathtt{t}-\Delta\mathtt{t}, \mathtt{t}]$. These vectors can be referred to a three-dimensional space, where $\boldsymbol{\mathfrak{a}}$ and $\boldsymbol{\mathfrak{x}}$ are applied to the origin of the reference system, $\boldsymbol{\mathfrak{s}}$ is applied to the destination of $\boldsymbol{\mathfrak{a}}$ and has the same destination of $\boldsymbol{\mathfrak{x}}$.

From $\boldsymbol{\mathfrak{a}}(\boldsymbol{\mathfrak{x}})$ and $\boldsymbol{\mathfrak{a}} \equiv \mathtt{a}_i$ we deduce, based on (19),

$$\mathtt{da}_i = \frac{\partial \mathtt{a}_i}{\partial \boldsymbol{\mathfrak{x}}} \cdot \mathtt{d}\boldsymbol{\mathfrak{x}} \qquad \text{i.e.} \qquad \mathtt{d}\boldsymbol{\mathfrak{a}} = \frac{\partial \boldsymbol{\mathfrak{a}}}{\partial \boldsymbol{\mathfrak{x}}} \cdot \mathtt{d}\boldsymbol{\mathfrak{x}}$$

of which

$$\frac{\partial \boldsymbol{\mathfrak{a}}}{\partial \boldsymbol{\mathfrak{x}}} \equiv \frac{\partial \mathtt{a}_i}{\partial \mathtt{x}_j} \qquad \text{i.e.} \qquad \mathtt{da}_i = \sum_{j=1}^{3} \frac{\partial \mathtt{a}_i}{\partial \mathtt{x}_j} \mathtt{dx}_j,$$

and analogously

$$d\mathbf{x} = \frac{\partial \mathbf{x}}{\partial \mathbf{a}} \cdot d\mathbf{a} \qquad \text{of which} \qquad \frac{\partial \mathbf{x}}{\partial \mathbf{a}} \equiv \frac{\partial x_i}{\partial a_j}.$$

A deformation is homogeneous if is constant the deformation gradient $\partial\mathbf{x}/\partial\mathbf{a}$. By specifying a well-known theorem of Mathematical Analysis (e.g. (2.4.6.2) of [22]), we have

$$\iiint_{\mathfrak{v}_t} f(\mathbf{x})\, d\mathfrak{v}_t = \iiint_{\mathfrak{v}_{t-\Delta t}} f(\mathbf{x}(\mathbf{a}))|\det[\partial\mathbf{x}/\partial\mathbf{a}]|\, d\mathfrak{v}_{t-\Delta t}$$

of which

$$d\mathfrak{v}_t \equiv \delta\mathfrak{v} \equiv \prod_{i=1}^{3} dx_i, \qquad d\mathfrak{v}_{t-\Delta t} \equiv \delta\mathfrak{v}_{t-\Delta t} \equiv \prod_{i=1}^{3} da_i,$$

where $[\partial\mathbf{x}/\partial\mathbf{a}]$ is the Jacobian matrix of $\mathbf{x}(\mathbf{a})$, and from which follows

$$\frac{\delta\mathfrak{v}_t}{\delta\mathfrak{v}_{t-\Delta t}} = |\det[\partial\mathbf{x}/\partial\mathbf{a}]|$$

From: $\mathbf{x} = \mathbf{a} + \mathbf{s}$, $\mathbf{s} = \mathbf{s}(\mathbf{a})$; (6), Ï$\langle a_i; i = 1, 3\rangle$; follows

$$\frac{\partial x_i}{\partial a_j} = \frac{\partial a_i}{\partial a_j} + \frac{\partial s_i}{\partial a_j} = \frac{\partial s_i}{\partial a_j} + \delta_{ij} \tag{46}$$

From: Æ⟨d$\boldsymbol{x}$, d$\boldsymbol{\alpha}$ // d$\mathbf{x}$ // (25)⟩; (6); (19); þ; (46); follows

$$
\begin{aligned}
|d\boldsymbol{x}|^2 - |d\boldsymbol{\alpha}|^2 &= \\
\sum_{h=1}^{3} dx_h dx_h - \sum_{i=1}^{3} da_i da_i &= \\
\sum_{k=1}^{3}\sum_{h=1}^{3} \delta_{hk} dx_h dx_k - \sum_{j=1}^{3}\sum_{i=1}^{3} \delta_{ij} da_i da_j &= \\
\sum_{j=1}^{3}\sum_{i=1}^{3}\left(\sum_{h=1}^{3}\sum_{k=1}^{3} \delta_{hk}\frac{\partial x_h}{\partial a_i}\frac{\partial x_k}{\partial a_j} - \delta_{ij}\right) da_i da_j &= \\
\sum_{j=1}^{3}\sum_{i=1}^{3}\left(\sum_{h=1}^{3} \frac{\partial x_h}{\partial a_i}\frac{\partial x_h}{\partial a_j} - \delta_{ij}\right) da_i da_j &= \\
\sum_{j=1}^{3}\sum_{i=1}^{3}\left(\sum_{h=1}^{3}\left(\frac{\partial s_h}{\partial a_i} + \delta_{hi}\right)\left(\frac{\partial s_h}{\partial a_j} + \delta_{hj}\right) - \delta_{ij}\right) da_i da_j &= \qquad (47) \\
\sum_{j=1}^{3}\sum_{i=1}^{3}\left(\sum_{h=1}^{3}\left(\frac{\partial s_h}{\partial a_i}\frac{\partial s_h}{\partial a_j} + \delta_{hi}\frac{\partial s_h}{\partial a_j} + \delta_{hj}\frac{\partial s_h}{\partial a_i} + \delta_{hi}\delta_{hj}\right) - \delta_{ij}\right) da_i da_j &= \\
\sum_{j=1}^{3}\sum_{i=1}^{3}\left(\sum_{h=1}^{3}\frac{\partial s_h}{\partial a_i}\frac{\partial s_h}{\partial a_j} + \sum_{h=1}^{3}\delta_{hi}\frac{\partial s_h}{\partial a_j} + \sum_{h=1}^{3}\delta_{hj}\frac{\partial s_h}{\partial a_i} + \sum_{h=1}^{3}\delta_{hi}\delta_{hj} - \delta_{ij}\right) da_i da_j &= \\
\sum_{j=1}^{3}\sum_{i=1}^{3}\left(\frac{\partial s_i}{\partial a_j} + \frac{\partial s_j}{\partial a_i} + \sum_{h=1}^{3}\frac{\partial s_h}{\partial a_i}\frac{\partial s_h}{\partial a_j}\right) da_i da_j &= \\
\sum_{j=1}^{3}\sum_{i=1}^{3} 2E_{ij} da_i da_j &= \\
d\boldsymbol{\alpha}\cdot 2\mathbf{E}\cdot d\boldsymbol{\alpha}
\end{aligned}
$$

where $\mathbf{E}$ is the deformation tensor of Green and St. Venant, expressed by

$$\mathbf{E} \equiv E_{ij} = \frac{1}{2}\left(\frac{\partial s_i}{\partial a_j} + \frac{\partial s_j}{\partial a_i} + \sum_{h=1}^{3}\frac{\partial s_h}{\partial a_i}\frac{\partial s_h}{\partial a_j}\right)$$

Analogously to (47) but with $\mathbf{s}(\boldsymbol{x})$ instead of $\mathbf{s}(\boldsymbol{\alpha})$, we deduce

$$|d\boldsymbol{x}|^2 - |d\boldsymbol{\alpha}|^2 = d\boldsymbol{x}\cdot 2\mathbf{e}\cdot d\boldsymbol{x} \qquad (48)$$

where $\mathbf{e}$ is the deformation tensor of Cauchy and Almansi, expressed by

$$\mathbf{e} \equiv e_{ij} = \frac{1}{2}\left(\frac{\partial s_i}{\partial x_j} + \frac{\partial s_j}{\partial x_i} - \sum_{h=1}^{3}\frac{\partial s_h}{\partial x_i}\frac{\partial s_h}{\partial x_j}\right)$$

and approximated by the infinitesimal deformation tensor of Cauchy $\boldsymbol{\varepsilon}$ of which

$$\boldsymbol{\varepsilon} \equiv \varepsilon_{ij} = \frac{1}{2}\left(\frac{\partial s_i}{\partial x_j} + \frac{\partial s_j}{\partial x_i}\right) \equiv \frac{\nabla \mathfrak{s} + (\nabla \mathfrak{s})^T}{2} \tag{49}$$

meaning that $(\nabla \mathfrak{s})^T$ is obtained exchanging the position of the subscripts in $\partial s_i/\partial x_j$ (and likewise for the gradient of any vector).

The $\mathbf{E}$ and $\mathbf{e}$ are said Lagrangian and Eulerian inasmuch inherent the places $\mathcal{V}_{t-\Delta t}$ and $\mathcal{V}_t$ of the matter that we consider respectively undeformed and deformed.

In coherence with the discretion $\mathcal{D}$ (in section 2.2), in $t - \Delta t$ we can immediately establish the convenient $\delta v \equiv \delta\mathfrak{v}$, but the result of the deformation of $\delta\mathfrak{v}$ between $t - \Delta t$ and t remains generally unknowable, because in t its boundary is unknown and in this context we can not take advantage of $\mathcal{D}$ to characterize it sufficiently without invalidate precisely the sought-after knowledge. This exclusion of $\mathcal{D}$ is due to the not being $\delta\mathfrak{v}_t$ infinitesimal with respect to th $\delta\mathfrak{v}_{t-\Delta t}$ with which it must be compared. Indeed in absence of further conditions, the information of shape deducible from $\partial\mathbf{x}/\partial\mathbf{a}$, as well as from $|d\mathbf{x}|^2 - |d\mathbf{a}|^2$, they can relate only straight segments, but when in absence of $\mathcal{D}$ the boundary of $\delta\mathfrak{v}$ is not necessarily constituted by plane faces delimited by such a type of segments.

However, if such deformation is infinitesimal i.e. negligible as constituted by infinitesimals of higher order, we can establish that $\delta\mathfrak{v}$ is a cube also in t, $|d\mathbf{a}|$ and $|d\mathbf{x}|$ are (on the basis of Æ⟨d$\mathbf{a}$, d$\mathbf{x}$ // dx // (25)⟩ the lengths of the diagonal of $\delta\mathfrak{v}$ in the two instants, and $|d\mathbf{x}|^2 - |d\mathbf{a}|^2$ describes the result of the deformation of $\delta\mathfrak{v}$ so exhaustively well that it can be considered as an its measure (we note that $d\mathbf{a}$ can also be considered as a generic infinitesimal vector, so it and $d\mathbf{x}$ are not necessarily the said diagonals).

This suitability of $|d\mathbf{x}|^2 - |d\mathbf{a}|^2$, if we can admit that $\delta\mathfrak{v}$ is a cube in both instants $t-\Delta t$ and t, it can be confirmed by a mental experiment approximable with a computer simulation where infinitesimals and infinities are constituted by quantities sufficiently small and large, and consisting, meaning Æ⟨$ş_c$ // ş⟩ and $ş_t$ as ş at the instant t, in introducing

$$\mathfrak{v}_{t-\Delta t} \equiv \{\delta\mathfrak{v}_{t-\Delta t,c}; c = 1, \infty\}$$

and then in deducing $\mathfrak{v}_t$ by substituting each $\delta\mathfrak{v}_{t-\Delta t,c}$ of diagonal $d\mathbf{a}_c$ with the correspondent $\delta\mathfrak{v}_{tc}$ of diagonal $d\mathbf{x}_c$.

This, (47) and (48) imply that the infinitesimality of the deformation of $\delta\mathfrak{v}$ lets describe its result by mean of $\mathbf{E}$ or $\mathbf{e}$ (not rarely some authors deduce these two tensors suitable to describe the results of finite deformations and so without necessity of said infinitesimality). In the following we consider only $\mathbf{e}$ because we are interested to functions of type $f(\underline{ж})$, but being analogous the treatment of $\mathbf{E}$ for functions of type $f(\underline{x}, t - \Delta t)$.

Indeed, subsisting such infinitesimality (and so negligibility), the description with $\mathbf{e}$ is confirmed by the following physical meanings: $(\partial s_i/\partial x_i)dx_i$ is a translation of $\delta\mathfrak{A}_{+i}$ with respect to $\delta\mathfrak{A}_i$ in the i-th direction, the distance between $\delta\mathfrak{A}_i$ and $\delta\mathfrak{A}_{+i}$ in $t-\Delta t$ and t is $dx_i-(\partial s_i/\partial x_i)dx_i$ and dx_i and so $\partial s_i/\partial x_i$ is a relative elongation per unit of length; moreover $(\partial s_i/\partial x_j)dx_j$ is a translation of $\delta\mathfrak{A}_{+j}$ with respect to $\delta\mathfrak{A}_j$ in the i-th direction, and so implies the change of shape indicated by $\partial s_i/\partial x_j$ as approximation of the angle $\arctan(\partial s_i/\partial x_j)$ (the smaller this angle is, the better the approximation is) between the planes of $\delta\mathfrak{A}_i$ before and after the deformation.

An infinitesimal deformation of $\delta\mathfrak{v}$ is equivalent to the infinitesimality of $\nabla\mathbf{s}$ i.e. $\partial s_i/\partial x_j$, and this condition subsists if is infinitesimal $\mathbf{s}$ i.e. s_i.

Therefore the infinitesimality of the deformation of $\delta\mathfrak{v}$ is guaranteed by an $\mathbf{s}$ infinitesimal, when though in the material world every displacement is finite as every other physical quantity.

This means that $\mathbf{e}$ cannot exactly describe the result of the deformation in a point $\underline{ж}$. However an infinitesimal $\mathbf{s}$, that is equivalent to a $d\mathbf{x}_c$, implies the disappearance of the distinction between $\mathbf{x}$ and $\mathbf{a}$ as well as that the derivatives present in the expressions of $\mathbf{E}$ and $\mathbf{e}$ are also infinitesimal whereby their products are negligible as infinitesimals of higher order, and so gives rise to $\mathbf{E} = \mathbf{e} = \boldsymbol{\varepsilon}$. Therefore $\boldsymbol{\varepsilon}$ can describe adequately a deformation as long as it corresponds to an $|\mathbf{s}|$ enough small (for this reason $\boldsymbol{\varepsilon}$ is said infinitesimal).

From: $\mathbf{x} = \mathbf{a}+\mathbf{s}$; $\mathbf{x} = \mathbf{x}_c$ due to Æ$\langle \mathbf{x} \,//\, f \,//\, (31)\rangle$, $\mathbf{a} = \mathbf{x}_c(t - \Delta t)$; (33), equality of the right and left derivatives due to

$$Æ\Big\langle \mathbf{x}_c(t) \in \mathbf{C}^0(t) \,//\, f(\underline{x}) \in \mathbf{C}^0(\underline{x}_0) \,//\, \text{section 1.3.1}\Big\rangle;$$

follows

$$\lim_{\Delta t\to 0}\frac{\mathbf{s}}{\Delta t} = \lim_{\Delta t\to 0}\frac{\mathbf{x}-\mathbf{a}}{\Delta t} = \lim_{\Delta t\to 0}\frac{\mathbf{x}_c(t)-\mathbf{x}_c(t-\Delta t)}{\Delta t} = \frac{d\mathbf{x}_c}{dt} = \mathbf{w}(\underline{ж}) \qquad (50)$$

The substitution of $\mathbf{s}$ with the displacement per time unit $d\mathbf{x}_c/dt$ and (50) imply that the deformation tensors $\mathbf{e}$ and $\boldsymbol{\varepsilon}$ (more pertinent to Solid

Mechanics) they become the rate strain tensors $\dot{\mathbf{e}}$ and $\dot{\boldsymbol{\varepsilon}}$ (more pertinent to Fluid Mechanics) of which

$$\dot{\mathbf{e}} \equiv \dot{e}_{ij} = \frac{1}{2}\left(\frac{\partial w_i}{\partial x_j} + \frac{\partial w_j}{\partial x_i} - \sum_{h=1}^{3} \frac{\partial w_h}{\partial x_i}\frac{\partial w_h}{\partial x_j}\right)$$

$$\dot{\boldsymbol{\varepsilon}} \equiv \dot{\varepsilon}_{ij} = \frac{1}{2}\left(\frac{\partial w_i}{\partial x_j} + \frac{\partial w_j}{\partial x_i}\right) \equiv \frac{\nabla \mathbf{w} + (\nabla \mathbf{w})^T}{2} \tag{51}$$

The being $\dot{\mathbf{e}}$ inherent the infinitesimal displacement $d\boldsymbol{\mathfrak{x}}_c$ implies that the deformation of $\delta\mathfrak{v}$ is infinitesimal and so that $\dot{\mathbf{e}}$ (unlike $\mathbf{e}$ inherent the finite $\boldsymbol{\mathfrak{s}}$) is an exact description of the strain rate in the point $\underline{ж}$. Instead this is not for $\dot{\boldsymbol{\varepsilon}}$ that approximates $\dot{\mathbf{e}}$ better to decrease of $|\mathbf{w}|$ but completely (i.e. verifying $\dot{\boldsymbol{\varepsilon}} = \dot{\mathbf{e}}$) only for values of this infinitesimal and so nonexistent in the material world.

Chapter 6

Forces and stresses

Regarding the following it is appropriate to remember the equivalence whereby a volume is replaceable by the PdM in it contained.

A force is imposed by a volume to another that undergoes it and is represented as a vector applied to a point of the second. Is implicit that the forces exist in the same instant, are of a same type and are undergone by a certain volume.

A force is mobile or immobile as the point where it is applied. Therefore the forces undergone by $\delta\mathfrak{v}$ and $\delta\mathrm{V}$ are respectively mobile and immobile, coherently with (17) and continuity of every function whereby we consider not t when $\delta\mathrm{V} \equiv \delta\mathfrak{v}$ but its neighborhoods when generally $\delta\mathrm{V} \not\equiv \delta\mathfrak{v}$. This distinction is not usual since in the technical literature is usually implicit that the forces are mobile and as such undergone by every PdM.

A force is internal or external respectively if is imposed by a part or by the surroundings of the volume the undergoes it. An infinitesimal volume undergoes only external forces, because are not considered its divisions in several volumes.

With reference to the distinction between real and inertial forces (the inertial are also named apparent or fictitious, and based on the D'Alembert's principle we can treat them as external forces), in the following are considered only real (i.e. not inertial) forces.

In this context of Continuum Mechanics a force is of surface (i.e. contact) or of volume (i.e. body). A contact force is applied to the boundary of a volume and is correlated to thermodynamic pressure or deformation of such volume. Are contact forces of second type those of friction, that are named viscous if regard fluids. Are body forces the gravitational, electric and magnetic field forces (as well as those inertial).

The application point of a force is considered as an element of a surface or of a volume in the respective cases of the homonym types. An external contact force is applied to a point of the external surface of the volume that undergoes it.

A vector sum of forces is named resultant and its application point, even having to be among those of the volume that undergoes the forces, is undetermined inasmuch they are different those of such forces. We mean this in saying that a volume undergoes a resultant.

The Newton's third law, or third principle of Dynamics or principle of action and reaction, affirms that for each force imposed by a PdM to another, the second imposes to the first a force of equal direction and magnitude, and of opposite sense. Therefore the forces exist in couples whose elements are equal and opposite vectors.

The being each force internal or external implies that a resultant is equal to the sum of two respective resultants of internal and external forces. However the first is deducible null by the Newton's third law. Thus a resultant is equal to the resultant of only external forces.

We mean $\delta\mathsf{V} \equiv \delta\mathfrak{v} \vee \delta\mathtt{V}$ whereby $\delta\mathsf{V}$ is the infinitesimal cube identified by $\underline{\mathsf{X}}$, of which $\mathsf{X} \equiv \mathfrak{x} \vee \mathtt{x}$, as that of its eight vertices that has minimum the value of every coordinate, whose external surface $\underline{\delta\mathsf{A}}$ is constituted by the union of the $\{\delta\mathsf{A}_{\mathrm{i}}, \delta\mathsf{A}_{+\mathrm{i}}; \mathrm{i} = 1, 3\}$ of which $\delta\mathsf{A} \equiv \delta\mathfrak{A} \vee \delta\mathtt{A}$ and whose $\delta\mathsf{A}_{+\mathrm{i}}$ constitutes also the boundary and is part of the infinitesimal cube $\delta\mathsf{V}_{+\mathrm{i}}$ adjacent to $\delta\mathsf{V}$ in the sense of X_{i} increasing. These $\{\delta\mathsf{A}_{\mathrm{i}}, \delta\mathsf{A}_{+\mathrm{i}}\}$ are distant $\mathrm{d}\mathsf{X}_{\mathrm{i}}$ and have the respective versors $\{-\hat{\mathbf{e}}_{\mathrm{i}}, \hat{\mathbf{e}}_{\mathrm{i}}\}$ of which $\hat{\mathbf{e}} \equiv \hat{e} \vee \hat{\mathbf{e}}$.

We name $\mathbf{F}$ and $\delta\mathbf{F}$, of which $\delta\mathbf{F} = \mathbf{f}\delta\mathsf{V}$, two resultants (of a same type of forces) undergone by the respective $\mathfrak{v}$ and $\delta\mathsf{V}$. In coherence with the being expressed by functions of $\underline{\text{ж}}$ and $\underline{\mathfrak{x}}$ physical quantities that describe the respective $\mathtt{c}$ and $\mathfrak{c}$, $\mathbf{f}$ is function of $\underline{\text{ж}}$ (and not of $\underline{\mathfrak{x}}$) if $\delta\mathsf{V} \equiv \delta\mathtt{V}$ or of $\underline{\mathfrak{x}}$ (and so, for Æ$\langle \mathbf{f} \,/\!/\, f \,/\!/\, (31)\rangle$, also of $\underline{\text{ж}}$) if $\delta\mathsf{V} \equiv \delta\mathfrak{v}$, consequently $\mathbf{f}$ can be in every case expressed by a function of $\underline{\text{ж}}$.

From: discretization of $\mathfrak{v}$ analogous to that of (43);

$$\text{Æ}\langle \mathbf{f}, \underline{\mathtt{x}}, \mathtt{t}, \delta\mathsf{V}, \mathfrak{v} \,/\!/\, g, \underline{\mathtt{x}}, \underline{\mathtt{e}}, \delta\mathtt{V}, \mathtt{V} \,/\!/\, (27)\rangle;$$

follows

$$\mathbf{F} = \sum_{\mathrm{c}=1}^{\infty} \mathbf{f}(\underline{\mathtt{x}}_{\mathrm{c}}, \mathtt{t})\delta\mathsf{V}_{\mathrm{c}} \equiv \iiint_{\mathfrak{v}} \mathbf{f}\,\mathrm{d}\mathfrak{v} \equiv \iiint_{\mathfrak{v}_{\mathtt{t}}} \mathbf{f}(\underline{\mathtt{x}}, \mathtt{t})\,\mathrm{d}\mathfrak{v} \tag{52}$$

whose last member has the aim of highlight $\mathbf{F}(\mathtt{t})$.

6.1 Stress vector

We name $\Delta\mathbf{F}_{\mathfrak{s}}$ the resultant of certain external forces of contact undergone by $\mathfrak{v}$ on a portion $\Delta\mathfrak{s}$ of its external surface $\mathfrak{s}$. In the limit as $\Delta\mathfrak{s} \to 0$, all the effects, that might have the alone $\Delta\mathbf{F}_{\mathfrak{s}}$ applied to the centroid of $\Delta\mathfrak{s}$, are the same of the inherent set of forces inasmuch the moment of every couples of such forces become negligible as infinitesimal of higher order. Therefore, by meaning

$$\delta\mathfrak{s} \equiv \lim_{\Delta\mathfrak{s}\to 0} \Delta\mathfrak{s}, \qquad \delta\mathbf{F}_{\mathfrak{s}} \equiv \lim_{\Delta\mathfrak{s}\to 0} \Delta\mathbf{F}_{\mathfrak{s}}, \qquad \mathbf{s}_{\mathfrak{s}} \equiv \lim_{\Delta\mathfrak{s}\to 0} \frac{\Delta\mathbf{F}_{\mathfrak{s}}}{\Delta\mathfrak{s}} = \frac{\delta\mathbf{F}_{\mathfrak{s}}}{\delta\mathfrak{s}}, \tag{53}$$

external forces of contact, undergone by $\mathfrak{v}$ on the infinitesimal areola $\delta\mathfrak{s}$ of versor $\hat{\mathbf{e}}_{\mathfrak{s}}$, are equivalent to their resultant $\delta\mathbf{F}_{\mathfrak{s}}$, are implicitly substituted by this intended as contact force applied in the centroid of $\delta\mathfrak{s}$, and $\mathbf{s}_{\mathfrak{s}}$ is their stress (or traction) vector that has magnitude of force per surface unit.

This and the Newton's third law entail, besides $\delta\mathbf{F}_{\mathfrak{s}}$ and $\mathbf{s}_{\mathfrak{s}}$ undergone by $\mathfrak{v}$ and imposed on $\delta\mathfrak{s}$ by the surroundings, $-\delta\mathbf{F}_{\mathfrak{s}}$ and $-\mathbf{s}_{\mathfrak{s}}$ imposed by $\mathfrak{v}$ and undergone on $\delta\mathfrak{s}$ by the surroundings.

A stress vector is named normal or tangential if is such with respect to the inherent surface. A tangential stress is also named shear stress.

How much just now of $\mathfrak{v}$ and $\mathfrak{s}$ can be referred even to $\delta\mathsf{V}$ and $\underline{\delta\mathsf{A}}$. Introducing so

$$\mathbf{s}_{\mathrm{i}} = \frac{\delta\mathbf{F}_{\mathrm{i}}}{\delta\mathsf{A}_{\mathrm{i}}} \qquad \text{of which} \qquad Æ\langle\delta\mathsf{A}_{\mathrm{i}}, \delta\mathbf{F}_{\mathrm{i}}, \mathbf{s}_{\mathrm{i}} \,/\!/\, \delta\mathfrak{s}, \delta\mathbf{F}_{\mathfrak{s}}, \mathfrak{s}\rangle$$

with the resultant $\delta\mathbf{F}_{\mathrm{i}}$ and the stress vector $\mathbf{s}_{\mathrm{i}}$ undergone by $\delta\mathsf{V}$ on $\delta\mathsf{A}_{\mathrm{i}}$ and imposed by the infinitesimal cube $\delta\mathsf{V}_{-\mathrm{i}}$ with which $\delta\mathsf{V}$ is in contact in the sense of $\mathrm{x_i}$ decreasing, we have (as well as for $\mathbf{f}$) $\mathbf{s}_{\mathrm{i}}(\underline{Ж})$ and

$$\mathbf{s}_{+\mathrm{i}} = \mathbf{s}_{\mathrm{i}}((\mathrm{x}_i + \delta_{\mathrm{i}i}\mathrm{dx}_i; i = 1, 3), \mathrm{t}) \equiv \mathbf{s}_{\mathrm{i}}(\underline{Ж}) + \frac{\partial\mathbf{s}_{\mathrm{i}}(\underline{Ж})}{\partial\mathrm{x_i}}\mathrm{dx_i} \tag{54}$$

with $\mathbf{s}_{+\mathrm{i}}$ the stress vector imposed by $\delta\mathsf{V}$ and undergone by $\delta\mathsf{V}_{+\mathrm{i}}$ on $\delta\mathsf{A}_{+\mathrm{i}}$, and the $\mathbf{s}_{\mathrm{i}}^{+} = -\mathbf{s}_{+\mathrm{i}}$ due to the Newton's third law with $\mathbf{s}_{\mathrm{i}}^{+}$ imposed by $\delta\mathsf{V}_{+\mathrm{i}}$ and undergone by $\delta\mathsf{V}$ on $\delta\mathsf{A}_{+\mathrm{i}}$.

The $\mathbf{s}_{\mathrm{i}}^{+} = -\mathbf{s}_{+\mathrm{i}}$, whose $\mathbf{s}_{\mathrm{i}}^{+}$ and $\mathbf{s}_{+\mathrm{i}}$ both are applied on $\delta\mathsf{A}_{+\mathrm{i}}$, implies null the resultant of contact forces applied on each contact square between two of the $\{\delta\mathsf{V}_{\mathrm{c}}; \mathrm{c} = 1, \infty\}$ that constitute $\mathfrak{v}$ and so confirms the nullity of the resultant of contact forces internal to $\mathfrak{v}$.

6.2 Stress tensor

Introducing $\mathbf{s}_i \equiv \tau_{ij}$ as specification of (1) (and analogously $\mathbf{s}_{+i} \equiv \tau_{+ij}$, $\mathbf{s}_i^+ \equiv \tau_{ij}^+$), we have the Cauchy stress tensor $\boldsymbol{\tau}$ defined by $\boldsymbol{\tau} \equiv \tau_{ij}$ and undergone by $\delta \mathsf{V}$ on the $(\delta \mathsf{A}_i; i = 1, 3)$ being $\tau_{ij}\hat{\mathbf{e}}_j$ the stress vector on $\delta \mathsf{A}_i$ and in the j-th direction.

A component τ_{ij} is named normal if $i = j$ or tangential if $i \neq j$, because is such $\tau_{ij}\hat{\mathbf{e}}_j$ with respect to $\delta \mathsf{A}_i$ (and analogously τ_{+ij} and τ_{ij}^+).

6.2.1 Default stress state

A stress vector is tensile (of tension) or compressive (of compression) if is directed towards the outside or inside of the volume that undergoes it.

To define, and to represent graphically as is particularly useful when we treat vectors, the default stress vectors undergone by volumes, it is necessary chose them conventionally tensile or compressive. These two cases define the two default stress states respectively named positive tension (or negative compression) and positive compression (or negative tension).

The negative or positive compression (the second might be preferable in case of fluids because these cannot undergo tensile stress vectors) give respectively rise to $\hat{\mathbf{e}}_{\S} \cdot \mathbf{s}_{\S} > 0$ or $\hat{\mathbf{e}}_{\S} \cdot \mathbf{s}_{\S} < 0$ inasmuch $\hat{\mathbf{e}}_{\S}$ is external.

From: (4), $\hat{\mathbf{e}}_i \equiv \delta_{ij}$, $\mathbf{s}_i \equiv \tau_{ij}$; (6); positive compression,

$$\text{Æ}\langle \delta \mathsf{A}_i, -\hat{\mathbf{e}}_i, \mathbf{s}_i \,/\!/\, \delta \S, \hat{\mathbf{e}}_{\S}, \mathbf{s}_{\S} \rangle;$$

follows

$$-\hat{\mathbf{e}}_i \cdot \mathbf{s}_i = -\sum_{k=1}^{3} \delta_{ik}\tau_{kj} = -\tau_{ij} < 0$$

from which $\tau_{ij} > 0$, and analogously we deduce $\tau_{ij} < 0$ in the case of negative compression.

Therefore in this context we chose the positive compression and $\tau_{ij} > 0$, because *vice versa* the negative compression and $\tau_{ij} < 0$ might impose the necessity to make exception to the positivity of the default value of every quantity.

As such $\tau_{ij} > 0$ we have also $\tau_{+ij} > 0$ that for $\mathbf{s}_i^+ = -\mathbf{s}_{+i}$ implies $\tau_{ij}^+ < 0$.

6.2.2 Symmetry of stress tensor

Excluding rotations and deformations of $\delta\mathbf{v}$ by means of the $\mathcal{D}$ of section 2.2, by the absence of rotation around the axis perpendicular to δA_k and passing for the centroid of $\delta\mathbf{v}$, and meaning $k \neq i \neq j$, it follows, based on the Newton's second law for the rotation ([90]) i.e. the Euler's second law for the rigid bodies ([6]), null the sum of the moments with respect to such axis of the forces

$$\tau_{ij}\delta A_i \hat{\mathbf{e}}_j, \qquad \tau_{ji}\delta A_j \hat{\mathbf{e}}_i, \qquad \tau^+_{ij}\delta A_{+i}\hat{\mathbf{e}}_j, \qquad \tau^+_{ji}\delta A_{+j}\hat{\mathbf{e}}_i.$$

This nullity, $\delta A_i = \delta A_{+i}$, $\tau_{ij} > 0$, and $\tau^+_{ij} < 0$ imply $\tau_{ij} = \tau_{ji}$ and $\tau^+_{ij} = \tau^+_{ji}$. This symmetry of $\boldsymbol{\tau}$ is usually deduced by the fundamental law of Continuum Mechanics constituted by the balance of the angular momentum (i.e. moment of momentum) with respect to a fixed point and substantially equivalent to that cited.

6.2.3 Cauchy's stress theorem

We name $\delta\mathbf{F}_C$ the resultant of contact forces of a certain type undergone by $\delta\mathbf{v}$. From: this;

$$\mathbf{s}^+_i = -\mathbf{s}_{+i},\ (54); \qquad \delta A_i = \delta A_{+i},\ \delta A_i dx_i = \delta\mathbf{v}; \qquad \mathbf{s}_i \equiv \tau_{ij}; \qquad \boldsymbol{\tau} \equiv \tau_{ij};$$

follows

$$\begin{gathered}\delta\mathbf{F}_C = \sum_{i=1}^{3}\left(\mathbf{s}_i\delta A_i + \mathbf{s}^+_i\delta A_{+i}\right) = \sum_{i=1}^{3}\left(\mathbf{s}_i\delta A_i - \left(\mathbf{s}_i + \frac{\partial\mathbf{s}_i}{\partial x_i}dx_i\right)\delta A_{+i}\right) = \\ -\delta\mathbf{v}\sum_{i=1}^{3}\frac{\partial\mathbf{s}_i}{\partial x_i} \equiv -\delta\mathbf{v}\sum_{i=1}^{3}\frac{\partial\tau_{ij}}{\partial x_i} \equiv -\delta\mathbf{v}\nabla\cdot\boldsymbol{\tau}\end{gathered} \tag{55}$$

whose sign "−" is eliminated from the last two members by adopting as default stress state the positive tension and by consider the default value of τ_{ij} positive as for every quantity.

Furthermore we name $\mathbf{F}_C$ the resultant of the contact forces (of the same type of $\delta\mathbf{F}_C$) undergone by $\mathfrak{v}$. From: $\delta\mathbf{F}_C = -\delta\mathbf{v}\nabla\cdot\boldsymbol{\tau}$ in (55),

$$\text{Æ}\langle \mathbf{F}_C, \delta\mathbf{F}_C = -\delta\mathbf{v}\nabla\cdot\boldsymbol{\tau} \mathbin{/\!/} \mathbf{F}, \delta\mathbf{F} = \mathbf{f}\delta\mathbf{v} \mathbin{/\!/} (52)\rangle;$$

Æ$\langle \boldsymbol{\tau}, \mathfrak{v}, \mathfrak{s} \,/\!/\, \mathbf{A}, \mathrm{V}, \mathrm{S} \,/\!/\, (36)\rangle$; follows

$$\mathbf{F}_C = -\iiint\limits_{\mathfrak{v}} \nabla \cdot \boldsymbol{\tau}\, \mathrm{d}\mathfrak{v} = -\iint\limits_{\mathfrak{s}} \hat{\mathbf{e}}_{\mathfrak{s}} \cdot \boldsymbol{\tau}\, \mathrm{d}\mathfrak{s} \tag{56}$$

We name $\mathbf{F}_{CE}$ the resultant of external contact forces (of the same type of $\mathbf{F}_C$) undergone by $\mathfrak{v}$. From: discretization of $\mathfrak{s}$ analogous to that of (43), $\delta\mathbf{F}_{\mathfrak{s}} = \mathbf{S}_{\mathfrak{s}}\delta\mathfrak{s}$ in (53);

$$Æ\langle \mathbf{S}_{\mathfrak{s}}, \underline{\mathrm{x}}, \mathrm{t}, \delta\mathfrak{s}, \mathfrak{s} \,/\!/\, g, \underline{\mathrm{x}}, \underline{\mathrm{e}}, \delta\mathrm{V}, \mathrm{V} \,/\!/\, (27)\rangle;$$

follows

$$\mathbf{F}_{CE} = \sum_{\mathrm{c}=1}^{\infty} \mathbf{S}_{\mathfrak{s}}(\underline{\mathrm{x}}_{\mathrm{c}}, \mathrm{t})\delta\mathfrak{s}_{\mathrm{c}} = \iint\limits_{\mathfrak{s}} \mathbf{S}_{\mathfrak{s}}\, \mathrm{d}\mathfrak{s} \tag{57}$$

From $\mathbf{F}_C = \mathbf{F}_{CE}$ (due to nullity of the resultant of internal forces), (56), (57) and genericity of $\mathfrak{s}$ follows the Cauchy's stress theorem

$$\mathbf{S}_{\mathfrak{s}} = -\hat{\mathbf{e}}_{\mathfrak{s}} \cdot \boldsymbol{\tau} \tag{58}$$

that is obtainable also by means of the homonymous tetrahedron.

6.3 Forces undergone by a body

A force, by summarizing the principal types, can be undergone or imposed, of contact or body, external or internal, immobile or mobile. Moreover a contact force is of thermodynamic pressure or deformation.

We name $\mathbf{F}_{ş}$ the resultant of all the forces undergone by $\mathfrak{v}$ but only of a certain type indicated by ş as a combination of the $\{\mathrm{T}, \mathrm{E}, \mathrm{C}, \mathrm{B}, \mathrm{M}, \mathrm{I}, \mathrm{P}, \mathrm{D}\}$ that specifying respectively all types, external, of contact, of body, mobile, immobile, of thermodynamic pressure, of deformation. We name $\boldsymbol{\tau}_{ş}$ a tensor that differs from $\boldsymbol{\tau}$ (whose type has not be specified) only for the specification indicated by ş. We admit that a $\boldsymbol{\tau}$ immobile or mobile is respectively undergone by $\delta\mathrm{V}$ or $\delta\mathfrak{v}$.

From: þ; þ; þ; þ; $\mathbf{F}_{\mathrm{BM}} = \mathbf{F}_{\mathrm{BME}}$ and $\mathbf{F}_{\mathrm{BI}} = \mathbf{F}_{\mathrm{BIE}}$ (due to nullity of the resultant of internal forces);

$$Æ\langle \mathbf{F}_{ş}, \mathbf{f}_{ş} \,/\!/\, \mathbf{F}, \mathbf{f} \,/\!/\, (52)\rangle; \qquad Æ\langle \mathbf{F}_{\mathrm{C}\ldots}, \mathbf{f}_{\mathrm{C}\ldots} \,/\!/\, \mathbf{F}_C, -\nabla \cdot \boldsymbol{\tau} \,/\!/\, (56)\rangle;$$

follows, for the resultant $\mathbf{F}_{\mathrm{T}}$ of all the forces undergone by $\mathfrak{v}$,

$$\begin{aligned}\mathbf{F}_{\mathrm{T}} = \mathbf{F}_{\mathrm{C}} + \mathbf{F}_{\mathrm{B}} = \mathbf{F}_{\mathrm{M}} + \mathbf{F}_{\mathrm{I}} = \mathbf{F}_{\mathrm{CM}} + \mathbf{F}_{\mathrm{CI}} + \mathbf{F}_{\mathrm{BM}} + \mathbf{F}_{\mathrm{BI}} = \\ \mathbf{F}_{\mathrm{CPM}} + \mathbf{F}_{\mathrm{CDM}} + \mathbf{F}_{\mathrm{CPI}} + \mathbf{F}_{\mathrm{CDI}} + \mathbf{F}_{\mathrm{BM}} + \mathbf{F}_{\mathrm{BI}} = \\ \mathbf{F}_{\mathrm{CPM}} + \mathbf{F}_{\mathrm{CDM}} + \mathbf{F}_{\mathrm{CPI}} + \mathbf{F}_{\mathrm{CDI}} + \mathbf{F}_{\mathrm{BME}} + \mathbf{F}_{\mathrm{BIE}} = \\ \iiint_{\mathfrak{v}} (\mathbf{f}_{\mathrm{CPM}} + \mathbf{f}_{\mathrm{CDM}} + \mathbf{f}_{\mathrm{CPI}} + \mathbf{f}_{\mathrm{CDI}} + \mathbf{f}_{\mathrm{BME}} + \mathbf{f}_{\mathrm{BIE}})\, d\mathfrak{v} = \\ \iiint_{\mathfrak{v}} (\mathbf{f}_{\mathrm{BME}} + \mathbf{f}_{\mathrm{BIE}} - \nabla\cdot\boldsymbol{\tau}_{\mathrm{PM}} - \nabla\cdot\boldsymbol{\tau}_{\mathrm{DM}} - \nabla\cdot\boldsymbol{\tau}_{\mathrm{PI}} - \nabla\cdot\boldsymbol{\tau}_{\mathrm{DI}})\, d\mathfrak{v}\end{aligned} \tag{59}$$

Introducing $\neg\mathcal{P}$ as the proposition true or false respectively if $\mathcal{P}$ is false or true,

$$\{\mathcal{P}_{\mathrm{A}} \Rightarrow \mathcal{P}_{\mathrm{B}}\} \equiv \{\neg\mathcal{P}_{\mathrm{B}} \Rightarrow \neg\mathcal{P}_{\mathrm{A}}\} \text{ (law of contraposition, e.g. in (3) of [3])},$$
$$\{\mathcal{P}_{\mathrm{A}} \Rightarrow \mathcal{P}_{\mathrm{B}}\} \wedge \{\mathcal{P}_{\mathrm{A}} \Leftarrow \mathcal{P}_{\mathrm{B}}\} \equiv \{\mathcal{P}_{\mathrm{A}} \equiv \mathcal{P}_{\mathrm{B}}\} \text{ ((2) of [3])},$$
$$\{\mathcal{P}_{\mathrm{A}} \equiv \mathcal{P}_{\mathrm{B}}\} \equiv \{\neg\mathcal{P}_{\mathrm{A}} \equiv \neg\mathcal{P}_{\mathrm{B}}\}, \qquad \mathcal{P}_{\mathrm{P}} \equiv \text{“}\boldsymbol{\tau} \equiv \boldsymbol{\tau}_{\mathrm{P}}\text{”} \equiv \neg\mathcal{P}_{\mathrm{D}}$$
$$\mathcal{P}_{\mathrm{D}} \equiv \text{“}\boldsymbol{\tau} \equiv \boldsymbol{\tau}_{\mathrm{D}}\text{”} \equiv \neg\mathcal{P}_{\mathrm{P}} \qquad \mathcal{P}_{\mathrm{I}} \equiv \text{“}\boldsymbol{\tau} \equiv \boldsymbol{\tau}_{\mathrm{I}}\text{”} \equiv \neg\mathcal{P}_{\mathrm{M}} \qquad \mathcal{P}_{\mathrm{M}} \equiv \text{“}\boldsymbol{\tau} \equiv \boldsymbol{\tau}_{\mathrm{M}}\text{”} \equiv \neg\mathcal{P}_{\mathrm{I}}$$

and admitting evident the $\mathcal{P}_{\mathrm{D}} \Rightarrow \mathcal{P}_{\mathrm{M}}$, we deduce as follows the $\mathcal{P}_{\mathrm{P}} \equiv \mathcal{P}_{\mathrm{I}}$ and $\mathcal{P}_{\mathrm{D}} \equiv \mathcal{P}_{\mathrm{M}}$ that affirm the inexistence of each $\boldsymbol{\tau}$ mobile of thermodynamic pressure or immobile of deformation, and so

$$\neg\exists\boldsymbol{\tau}_{\mathrm{PM}} \qquad\qquad \neg\exists\boldsymbol{\tau}_{\mathrm{DI}} \tag{60}$$

that imply respectively $\neg\exists\mathbf{F}_{\mathrm{CPM}}$ and $\neg\exists\mathbf{F}_{\mathrm{CDI}}$.

Indeed

$$\mathcal{P}_{\mathrm{P}} \Rightarrow \mathcal{P}_{\mathrm{M}} \quad \text{i.e.} \quad \neg\mathcal{P}_{\mathrm{M}} \Rightarrow \neg\mathcal{P}_{\mathrm{P}} \qquad \text{and} \qquad \mathcal{P}_{\mathrm{D}} \Rightarrow \mathcal{P}_{\mathrm{M}} \quad \text{i.e.} \quad \neg\mathcal{P}_{\mathrm{M}} \Rightarrow \neg\mathcal{P}_{\mathrm{D}}$$

give rise to $\neg\mathcal{P}_{\mathrm{M}} \Rightarrow \{\neg\mathcal{P}_{\mathrm{P}} \wedge \neg\mathcal{P}_{\mathrm{D}}\}$ that is false inasmuch equivalent to $\mathcal{P}_{\mathrm{I}} \Rightarrow \{\mathcal{P}_{\mathrm{P}} \wedge \mathcal{P}_{\mathrm{D}}\}$ whose second member is impossible because a force cannot be both of thermodynamic pressure and of deformation.

So we have a *demonstratio per absurdum* (e.g. section 2 of [3]) of the $\neg\{\mathcal{P}_{\mathrm{P}} \Rightarrow \mathcal{P}_{\mathrm{M}}\}$ (equivalent to $\neg\{\mathcal{P}_{\mathrm{I}} \Rightarrow \mathcal{P}_{\mathrm{D}}\}$ that, moreover, is not less evident of $\mathcal{P}_{\mathrm{D}} \Rightarrow \mathcal{P}_{\mathrm{M}}$).

The $\neg\{\mathcal{P}_{\mathrm{P}} \Rightarrow \mathcal{P}_{\mathrm{M}}\}$ implies $\mathcal{P}_{\mathrm{P}} \Rightarrow \neg\mathcal{P}_{\mathrm{M}}$ i.e. $\mathcal{P}_{\mathrm{M}} \Rightarrow \mathcal{P}_{\mathrm{D}}$. This and $\mathcal{P}_{\mathrm{D}} \Rightarrow \mathcal{P}_{\mathrm{M}}$ give rise to the searched $\mathcal{P}_{\mathrm{D}} \equiv \mathcal{P}_{\mathrm{M}}$.

From: þ; $\mathcal{P}_D \equiv \mathcal{P}_M$; follows

$$\mathcal{P}_P \equiv \neg\mathcal{P}_D \equiv \neg\mathcal{P}_M \equiv \mathcal{P}_I$$

and so the searched $\mathcal{P}_P \equiv \mathcal{P}_I$.

The $\neg\exists\boldsymbol{\tau}_{DI}$ is confirmed by the fact that $\boldsymbol{\tau}_{DI}$, inasmuch immobile, should be correlated to the deformation of $\delta\mathrm{v}$, when however this, in addition to having outer surface crossable by matter, is immobile, immutable and indeformable, being able therefore to exchanging matter with the surroundings without being able to have any deformation.

A pressure is a magnitude with sign of a normal stress vector; e.g. the pressure $\mathsf{p}_{\mathfrak{s}}$, that the surroundings of $\mathfrak{v}$ exercises on $\delta\mathfrak{s}$, is (given the current external versor and positive compression) the magnitude with sign of the stress vector $-\mathsf{p}\hat{\mathbf{e}}_{\mathfrak{s}}$ i.e. $\mathsf{p}_{\mathfrak{s}}$ is the component of $\mathbf{s}_{\mathfrak{s}}$ along $-\hat{\mathbf{e}}_{\mathfrak{s}}$ inasmuch

$$Æ\langle \mathbf{s}_{\mathfrak{s}}, -\hat{\mathbf{e}}_{\mathfrak{s}}, \mathsf{p}_{\mathfrak{s}} /\!\!/ \mathbf{a}, \hat{\mathbf{e}}_N, \hat{\mathbf{e}}_N \cdot \mathbf{a} /\!\!/ (12)\rangle.$$

The $\neg\exists\boldsymbol{\tau}_{PM}$ is confirmed by the fact that $\boldsymbol{\tau}_{PM}$, inasmuch mobile, should be correlated to the thermodynamic pressure of $\delta\mathfrak{v}$, when however the isotropy of this quantity, made immediately evident by that of density and temperature to which is linked by the equation of state, is incompatible with the three generally different pressures $\{\tau_{DMii}; i = 1, 3\}$ correlated to the deformation of $\delta\mathfrak{v}$, being τ_{DMii} that of $\tau_{DMii}\hat{\mathfrak{e}}_i$. In other words $\neg\exists\boldsymbol{\tau}_{PM}$ is coherent with the inexistence, of an only thermodynamic pressure in the infinitesimal $\delta\mathfrak{v}$, caused by the simultaneity of the three generally different pressures of deformation $\{\tau_{DMii}; i = 1, 3\}$ on the respective $\{\delta\mathfrak{A}_i; i = 1, 3\}$.

The necessity of distinguish between mobile and immobile forces, as well as between $\delta\mathfrak{v}$ and $\delta\mathrm{v}$ of which $\delta\mathrm{v} \equiv \delta\mathfrak{v}$ in t, is evidenced by having to consider, imposed by (17) and implicit continuity of every function, not the point t but limits approaching it.

Indeed in the limit as $t \to \mathrm{t}$ the surfaces $\underline{\delta\mathfrak{A}}$ and $\underline{\delta\mathrm{A}}$ they get closer and closer tending to overlap but remain distinct, and this induces to distinguish the respective contact forces undergone by $\delta\mathfrak{v}$ and $\delta\mathrm{v}$ as different quantities i.e. as properties of different objects consisting of distinct surfaces.

An analogous distinction, based on the fact that in the said limit also $\delta\mathfrak{v}$ and $\delta\mathrm{v}$ remain distinct analogously to $\underline{\delta\mathfrak{A}}$ and $\underline{\delta\mathrm{A}}$, might be adopted also between the respective body forces undergone by $\delta\mathfrak{v}$ and $\delta\mathrm{v}$, but is instead negated for the following reasons.

Based on (24) we have

$$\frac{\delta \mathsf{V}}{\delta \mathsf{A}_i} = \frac{\delta \mathsf{x}_i \delta \mathsf{x}_j \delta \mathsf{x}_k}{\delta \mathsf{x}_j \delta \mathsf{x}_k} = \delta \mathsf{x}_i = \lim_{\Delta x \to 0} \Delta x = 0 \tag{61}$$

The distinction among two sets $\underline{A}$ and $\underline{B}$ can be measured by $\check{\mathbf{D}}\langle \underline{A}, \underline{B} \rangle$ of which

$$\check{\mathbf{D}}\{\underline{A}, \underline{B}\} \equiv \{\underline{A} - \underline{B}\} + \langle \underline{B} - \underline{A} \rangle.$$

The being $\delta \mathsf{V}$ an infinitesimal of higher order with respect to $\underline{\delta \mathsf{A}}$ (as we deduce from (61)) implies that also $\check{\mathbf{D}}\langle \delta \mathfrak{v}, \delta \mathrm{V} \rangle$ is such with respect to $\check{\mathbf{D}}\langle \underline{\delta \mathfrak{A}}, \underline{\delta \mathrm{A}} \rangle$ and so the irrelevance of the distinction among $\delta \mathfrak{v}$ and $\delta \mathrm{V}$ when is considered that among $\underline{\delta \mathfrak{A}}$ and $\underline{\delta \mathrm{A}}$.

Furthermore, in fact, the distinction between contact forces undergone by $\delta \mathfrak{v}$ and $\delta \mathrm{V}$ is equivalent to that among the respective $\underline{\delta \mathfrak{A}}$ and $\underline{\delta \mathrm{A}}$, as well as the distinction among body forces undergone by $\delta \mathfrak{v}$ and $\delta \mathrm{V}$ is equivalent to that among such volumes.

Hence a distinction among body forces undergone by $\delta \mathfrak{v}$ and $\delta \mathrm{V}$, analogous to that just now accepted for the contact forces, is impeded by the said irrelevance of the distinction among such volumes with respect to that among their surfaces.

This is confirmed by considering that such two types of body forces not have, besides mobility and immobility (respectively due to the being undergone by $\delta \mathfrak{v}$ and $\delta \mathrm{V}$), further distinctive characteristics analogous to the aforementioned for the contact forces undergone by $\underline{\delta \mathfrak{A}}$ and $\underline{\delta \mathrm{A}}$.

This indistinction among body forces and the obvious usual existence of the mobile body forces induce to admit

$$\neg\exists \mathbf{F}_{BI} \qquad \neg\exists \mathbf{F}_{BIE} \tag{62}$$

How much just now for the forces generally is worth for each quantity, in the sense of the indistinction, and so uniqueness, of two quantities that differ for the being properties of $\delta \mathfrak{v}$ but not of $\underline{\delta \mathfrak{A}}$ or of $\delta \mathrm{V}$ but not of $\underline{\delta \mathrm{A}}$, and of the distinction, and so duplicity, of two quantities that differ for the being properties of $\underline{\delta \mathfrak{A}}$ or $\underline{\delta \mathrm{A}}$.

We note furthermore that (60) and (62) are based on infinitesimal schemes allowed by the implicit $\mathcal{D}$ of section 2.2.

The $\mathsf{P}(\underline{ж})$, $\mathsf{T}(\underline{ж})$ and $\rho(\underline{ж})$ respectively express the thermodynamic pressure P, the temperature T and the density ρ of the substance of which is composed the PdM C that occupies $\delta\mathrm{V}$. This three variables are those of the equation of state of such substance that, with the current positive compression, express P respectively positive or negative if is compressive or tensile, and being furthermore P isotropic also because are so ρ and T.

In coherence with this and the being a pressure the magnitude with sign of a normal stress vector, we have

$$Æ\langle \mathsf{P}\hat{\mathbf{e}}_\mathrm{i}, \delta\mathbb{A}_\mathrm{i}, -\hat{\mathbf{e}}_\mathrm{i}, \boldsymbol{\tau}_\mathrm{PI} \,/\!/\, \mathbf{s}_{\S}, \delta\S, \hat{\mathbf{e}}_{\S}, \boldsymbol{\tau} \,/\!/\, (58)\rangle$$

that gives rise to $\mathsf{P}\hat{\mathbf{e}}_\mathrm{i} = \hat{\mathbf{e}}_\mathrm{i} \cdot \boldsymbol{\tau}_\mathrm{PI}$ of which $\boldsymbol{\tau}_\mathrm{PI} \equiv \tau_\mathrm{PIij}$ with $\boldsymbol{\tau}_\mathrm{PI}$ the stress tensor undergone by $\delta\mathrm{V}$ on the $\{\delta\mathbb{A}_\mathrm{i}; \mathrm{i} = 1, 3\}$ and function of P.

Writing such $\mathsf{P}\hat{\mathbf{e}}_\mathrm{i} = \hat{\mathbf{e}}_\mathrm{i} \cdot \boldsymbol{\tau}_\mathrm{PI}$ in scalar terms by means of (3) and (4), and meaning (for (60)) $\boldsymbol{\tau}_\mathrm{PI} \equiv \boldsymbol{\tau}_\mathrm{P} \equiv \tau_\mathrm{Pij}$, we have

$$\mathsf{P}\delta_\mathrm{ij} = \sum_{\mathrm{h}=1}^{3} \delta_\mathrm{ih}\tau_\mathrm{Phj} = \tau_\mathrm{Pij}$$

i.e. $\boldsymbol{\tau}_\mathrm{P} = \mathsf{P}\boldsymbol{\delta}$ of which $\boldsymbol{\delta} \equiv \delta_\mathrm{ij}$.

From:

$$\boldsymbol{\tau}_\mathrm{P} = \mathsf{P}\boldsymbol{\delta}; \qquad (16); \qquad \nabla \cdot \boldsymbol{\delta} = \mathbf{0} \quad \text{of which} \quad \mathbf{0} \equiv 0; \qquad (4);$$

follows

$$\nabla \cdot \boldsymbol{\tau}_\mathrm{P} = \nabla \cdot \mathsf{P}\boldsymbol{\delta} = \mathsf{P}\nabla \cdot \boldsymbol{\delta} + \nabla\mathsf{P} \cdot \boldsymbol{\delta} = \nabla\mathsf{P} \cdot \boldsymbol{\delta} = \nabla\mathsf{P} \tag{63}$$

From:

$$Æ\langle \mathbf{F}_\mathrm{T}, \mathbf{f}_\mathrm{T} \,/\!/\, \mathbf{F}, \mathbf{f} \,/\!/\, (52)\rangle; \qquad (59), \quad (60), \quad (62), \quad (15);$$
$$\boldsymbol{\tau}_\mathrm{DM} \equiv \boldsymbol{\tau}_\mathrm{D}, \quad \boldsymbol{\tau}_\mathrm{PI} \equiv \boldsymbol{\tau}_\mathrm{P}, \quad \mathbf{f}_\mathrm{BME} \equiv \mathbf{f}_\mathrm{BE} \text{ (justified by (60) and (62))};$$

(63); follows

$$\begin{gathered}\mathbf{F}_\mathrm{T} = \iiint_{\mathfrak{v}} \mathbf{f}_\mathrm{T}\, d\mathfrak{v} = \iiint_{\mathfrak{v}} (\mathbf{f}_\mathrm{BME} - \nabla \cdot (\boldsymbol{\tau}_\mathrm{DM} + \boldsymbol{\tau}_\mathrm{PI}))\, d\mathfrak{v} = \\ \iiint_{\mathfrak{v}} (\mathbf{f}_\mathrm{BE} - \nabla \cdot (\boldsymbol{\tau}_\mathrm{D} + \boldsymbol{\tau}_\mathrm{P}))\, d\mathfrak{v} = \iiint_{\mathfrak{v}} (\mathbf{f}_\mathrm{BE} - \nabla\mathsf{P} - \nabla \cdot \boldsymbol{\tau}_\mathrm{D})\, d\mathfrak{v}\end{gathered} \tag{64}$$

of which

$$\text{Æ}\langle \mathbf{f}_{\text{BE}} \mathbin{/\!/} f \mathbin{/\!/} (31)\rangle \qquad \text{and} \qquad \text{Æ}\langle \delta\mathbf{F}_{\text{BE}}, \mathbf{f}_{\text{BE}}, \delta\mathfrak{v} \mathbin{/\!/} \delta\mathbf{F}, \mathbf{f}, \delta\mathsf{V} \mathbin{/\!/} (52)\rangle$$

with $\mathbf{f}_{\text{BE}}$ the resultant, per volume unit, of the body forces undergone by $\delta\mathfrak{v}$ and imposed by the surroundings of $\mathfrak{v}$, and where $\boldsymbol{\tau}_{\text{D}}$ (of which $\boldsymbol{\tau}_{\text{D}} \equiv \tau_{\text{Dij}}$) is the stress tensor undergone by $\delta\mathfrak{v}$ on the $\{\delta\mathfrak{A}_{\text{i}}; \text{i} = 1, 3\}$ and connected to the deformation of $\delta\mathfrak{v}$.

The correctness of (64) is confirmed by having obtained it by considering every distinguishable type of forces and without repetitions of the same addends.

In coherence with (64) and the being able to be only of deformation the tangential components of stress tensors, we introduce

$$\boldsymbol{\tau}_{\text{T}} = \boldsymbol{\tau}_{\text{D}} + \boldsymbol{\tau}_{\text{P}} = \boldsymbol{\tau}_{\text{D}} + \mathsf{P}\boldsymbol{\delta} = \boldsymbol{\mathcal{T}}_{\text{T}} + \overline{\mathsf{P}}\boldsymbol{\delta} \tag{65}$$

with $\boldsymbol{\mathcal{T}}_{\text{T}}$ the deviator tensor of which $\boldsymbol{\mathcal{T}}_{\text{T}} \equiv \mathcal{T}_{\text{Tij}}$ and that has null trace ($\sum_{\text{i}=1}^{3} \mathcal{T}_{\text{Tii}} = 0$) because $\overline{\mathsf{P}} = \sum_{\text{i}=1}^{3} \tau_{\text{Tii}}/3$. Such $\overline{\mathsf{P}}$, although it also has the name of mechanical pressure, it is though the mere average of the normal components of $\boldsymbol{\tau}_{\text{T}}$ and is thus a real pressure of the material world only if $\tau_{\text{Tii}} = \tau_{\text{Tjj}}$.

From: (64); $\nabla\mathsf{P} = \nabla \cdot \mathsf{P}\boldsymbol{\delta}$ in (63), (15); (65); $\mathbf{f}_{\hbar} = \mathbf{f}_{\text{BE}} - \nabla \cdot \boldsymbol{\tau}_{\text{D}}$; follows

$$\mathbf{f}_{\text{T}} = \mathbf{f}_{\text{BE}} - \nabla\mathsf{P} - \nabla \cdot \boldsymbol{\tau}_{\text{D}} = \mathbf{f}_{\text{BE}} - \nabla \cdot (\mathsf{P}\boldsymbol{\delta} + \boldsymbol{\tau}_{\text{D}}) = \mathbf{f}_{\text{BE}} - \nabla \cdot \boldsymbol{\tau}_{\text{T}} = \mathbf{f}_{\hbar} - \nabla\mathsf{P} \tag{66}$$

where, having $\delta\mathsf{V} \equiv \delta\mathfrak{v} \equiv \delta\mathsf{V} \cup \delta\mathfrak{v}$ in the implicit $\mathtt{t}$, $-\nabla\mathsf{P}\delta\mathsf{V}$ and $\mathbf{f}_{\hbar}\delta\mathfrak{v}$ are the respective resultants of all forces undergone by $\delta\mathsf{V}$ and $\delta\mathfrak{v}$, and instead $\nabla \cdot \boldsymbol{\tau}_{\text{T}}\delta\mathsf{V}$ and $\mathbf{f}_{\text{T}}\delta\mathsf{V}$ both are resultants of forces undergone by $\delta\mathsf{V}\cup\delta\mathfrak{v}$ because the resultant of a set of forces is the sum of the elements of such set and so the associativity of such a sum evidences as the resultant of the forces applied to the union of volumes is equal to the sum of the resultants inherent such volumes.

6.4 Stress constitutive equation

We say constitutive a relation (e.g. an equation of state) peculiar of a certain substance inasmuch identifies, characterizes and describes it being determined by means of specific experiments and/or sub-macroscopic models. A law and a constitutive relation both describe the material world, but one worth exactly for every substance and instead the other is pertinent only one substance idealized and schematized inasmuch by it described as approximation of the material reality.

The $\boldsymbol{\tau}_D$ is, together other variables that describe the deformation of $\delta\mathfrak{v}$, between the variables of a stress constitutive equation of the substance of which $\mathfrak{C}$ is composed.

The most known stress constitutive equations are the Hooke's linear elasticity law generalized for the isotropic substances i.e.

$$\boldsymbol{\tau}_T = 2\mu\boldsymbol{\varepsilon} + \lambda(\nabla\cdot\mathbf{s})\boldsymbol{\delta} \qquad \text{i.e.} \qquad \boldsymbol{\varepsilon} = \frac{(1+\nu)\boldsymbol{\tau}_T - \nu\sum_{i=1}^{3}\tau_{Tii}\boldsymbol{\delta}}{E} \tag{67}$$

where $\boldsymbol{\tau}_D$ is substituted by $\boldsymbol{\tau}_T$ on the basis of (65) and inasmuch there is no interest in distinguishing $\mathsf{P}\boldsymbol{\delta}$, whose $\boldsymbol{\varepsilon}$ is defined by (49), with μ and λ the Lamé's constants, E the Young's modulus of normal elasticity, ν the Poisson's coefficient of transverse contraction; and the equation of isotropic Newtonian fluids (this also evidently linear) that expresses the viscous stresses i.e. the first of the

$$\boldsymbol{\tau}_D = 2\mu\dot{\boldsymbol{\varepsilon}} + \lambda(\nabla\cdot\mathbf{w})\boldsymbol{\delta} \qquad\qquad \boldsymbol{\tau}_D = 2\mu\left(\dot{\boldsymbol{\varepsilon}} - \frac{(\nabla\cdot\mathbf{w})\boldsymbol{\delta}}{3}\right) \tag{68}$$

whose $\dot{\boldsymbol{\varepsilon}}$ is defined by (51), with μ and λ the viscosity coefficients, whose second follows from the first and the Stokes' hypothesis $\lambda = -2\mu/3$, but being generally λ an unknown in consequence of the impossibility of its experimental measures and of the indemonstrability of said hypothesis ([8, 10, 76]). Generally μ, λ, E, ν, μ, λ are functions of other variables such as T and P.

From: þ; (65); first of (68), (51); follows

$$\overline{\mathsf{P}} = \frac{\sum_{i=1}^{3}\tau_{Tii}}{3} = \frac{\sum_{i=1}^{3}(\tau_{Dii}+\mathsf{P})}{3} = k\nabla\cdot\mathbf{w} + \mathsf{P} \tag{69}$$

of which $k = \lambda + 2\mu/3$ with k the bulk viscosity coefficient, and that shows as $k = 0$ (which follows from the Stokes' hypothesis) and/or $\nabla\cdot\mathbf{w} = 0$ imply $\overline{\mathsf{P}} = \mathsf{P}$.

The second of (68) is equivalent to

$$\tau_{Dij} = 2\mu\left(\dot{\varepsilon}_{ij} - \frac{\nabla\cdot\mathbf{w}\delta_{ij}}{3}\right)$$

that, substituting $\dot{\varepsilon}_{ij}$ with its expression in (51) and $\nabla\cdot\mathbf{w}$ with the expression of the divergence of a vector, becomes

$$\tau_{Dij} = \mu\left(\frac{\partial \mathsf{w}_i}{\partial \mathsf{x}_j} + \frac{\partial \mathsf{w}_j}{\partial \mathsf{x}_i} - \frac{2}{3}\sum_{k=1}^{3}\delta_{ij}\frac{\partial \mathsf{w}_k}{\partial \mathsf{x}_k}\right)$$

and hence the

$$\tau_{Dij} = \mu \sum_{h=1}^{3} \sum_{k=1}^{3} \delta_{ijkh} \frac{\partial w_k}{\partial x_h} \tag{70}$$

of which

$$\delta_{ijkh} \equiv \delta_{ik}\delta_{jh} + \delta_{jk}\delta_{ih} - \frac{2}{3}\delta_{hk}\delta_{ij}$$

and that is another writing of the second of (68).

Chapter 7

Deduction of the New Law

7.1 Newton's second law

Every volume has a point as its center of mass. The position $\boldsymbol{x}_V$, of the center of mass $\underline{x}_V$ of V, is the vector applied to the origin of the three-dimensional reference system of $\boldsymbol{\mathcal{V}}$ and that has as destination $\underline{x}_V$ of which $\underline{x}_V \equiv (x_{Vi}; i = 1, 3)$. In this regard we have

$$\boldsymbol{x}_V = \frac{\iiint_V \rho \boldsymbol{x} \, dV}{M} \qquad \mathbf{w}_V = \frac{d\boldsymbol{x}_V}{dt} \tag{71}$$

of which $\boldsymbol{x}_V \equiv x_{Vi}$, (44), $\boldsymbol{x} \equiv x_i$, and with $\mathbf{w}_V$ the velocity of $\underline{x}_V$.

From:

$$\text{Æ}\langle \mathfrak{v}, \mathfrak{m} \,/\!/\, V, M \,/\!/\, (71) \rangle;$$

$$\text{constancy of } \mathfrak{m}, \quad \text{Æ}\langle \rho x_i \,/\!/\, f \,/\!/\, (42) \rangle, \quad \frac{\partial x_i}{\partial t} = 0 \quad \text{due to} \quad \text{Ï}\langle \underline{\text{ж}} \rangle;$$

$$\text{Æ}\langle x_i, \rho \mathbf{w} \,/\!/\, A, \mathbf{B} \,/\!/\, (16) \rangle; \qquad (45), \quad (14); \qquad \frac{\partial x_i}{\partial x_j} = \delta_{ij} \quad \text{due to} \quad \text{Ï}\langle \underline{\text{ж}} \rangle;$$

follows

$$\begin{gathered} \mathbf{w}_\mathfrak{v} \equiv \frac{d\left(\mathfrak{m}^{-1} \iiint_\mathfrak{v} \rho x_i \, d\mathfrak{v}\right)}{dt} = \mathfrak{m}^{-1} \iiint_\mathfrak{v} \left(x_i \frac{\partial \rho}{\partial t} + \nabla \cdot \rho x_i \mathbf{w} \right) d\mathfrak{v} = \\ \mathfrak{m}^{-1} \iiint_\mathfrak{v} \left(x_i \left(\frac{\partial \rho}{\partial t} + \nabla \cdot \rho \mathbf{w} \right) + \nabla x_i \cdot \rho \mathbf{w} \right) d\mathfrak{v} = \\ \mathfrak{m}^{-1} \iiint_\mathfrak{v} \rho \sum_{j=1}^{3} w_j \frac{\partial x_i}{\partial x_j} \, d\mathfrak{v} = \mathfrak{m}^{-1} \iiint_\mathfrak{v} \rho w_i \, d\mathfrak{v} \equiv \mathfrak{m}^{-1} \iiint_\mathfrak{v} \rho \mathbf{w} \, d\mathfrak{v} \end{gathered} \tag{72}$$

From: this;

$$Æ\langle \rho w_i \mathbin{//} f \mathbin{//} (42)\rangle, \quad Æ\langle w_i, \rho \mathbf{w} \mathbin{//} A, \mathbf{B} \mathbin{//} (16)\rangle;$$
$$(45), \quad \nabla w_i \cdot \rho \mathbf{w} = \rho \nabla w_i \cdot \mathbf{w};$$

follows

$$\frac{\mathrm{d}\mathfrak{m}\mathbf{w}_{\mathfrak{v}}}{\mathrm{d}t} \equiv \frac{\mathrm{d}\iiint_{\mathfrak{v}} \rho w_i \,\mathrm{d}\mathfrak{v}}{\mathrm{d}t} =$$
$$\iiint_{\mathfrak{v}} \left(w_i \left(\frac{\partial \rho}{\partial t} + \nabla \cdot \rho \mathbf{w} \right) + \rho \frac{\partial w_i}{\partial t} + \nabla w_i \cdot \rho \mathbf{w} \right) \mathrm{d}\mathfrak{v} = \tag{73}$$
$$\iiint_{\mathfrak{v}} \rho \left(\frac{\partial w_i}{\partial t} + \nabla w_i \cdot \mathbf{w} \right) \mathrm{d}\mathfrak{v} \equiv \iiint_{\mathfrak{v}} \rho \left(\frac{\partial \mathbf{w}}{\partial t} + \nabla \mathbf{w} \cdot \mathbf{w} \right) \mathrm{d}\mathfrak{v}$$

Analogously to this and considering (35), we also deduce

$$\frac{\mathrm{d}\iiint_{\mathfrak{v}} \rho f \,\mathrm{d}\mathfrak{v}}{\mathrm{d}t} = \iiint_{\mathfrak{v}} \rho \frac{\mathrm{d}f(\underline{\mathbf{x}}, t)}{\mathrm{d}t} \mathrm{d}\mathfrak{v}$$

The Newton's second law (or second principle of Dynamics), regardless of the debate on the uncertainties of the originating exposition in [92] (see [6, 19, 21, 36, 88]), is usually mentioned in relation to special points of $\mathcal{V}$ that are said material inasmuch endowed of respective finite masses. This law, for a material point $\mathcal{P}$ in which only the force $\boldsymbol{\mathcal{F}}$ is applied, that is endowed of mass $\mathcal{M}$ and moves with velocity $\boldsymbol{\mathcal{W}}$, has the most general formulation

$$\frac{\mathrm{d}\mathcal{M}\boldsymbol{\mathcal{W}}}{\mathrm{d}t} = \mathcal{M}\frac{\mathrm{d}\boldsymbol{\mathcal{W}}}{\mathrm{d}t} + \boldsymbol{\mathcal{W}}\frac{\mathrm{d}\mathcal{M}}{\mathrm{d}t} = \boldsymbol{\mathcal{F}} \tag{74}$$

that to describe the material world evidently needs to increase the said definitions of $\mathcal{P}$, $\mathcal{M}$ and $\boldsymbol{\mathcal{F}}$.

Considering to this respect the inexistence of points that might contain masses (a point has not extension) and the uniqueness of $\boldsymbol{\mathcal{F}}$, the only meaning of (74) that remains possible is based on the interpreting $\mathcal{P}$ as a point representative between those of a volume $\mathcal{V}$ of mass $\mathcal{M}$, that has $\boldsymbol{\mathcal{F}}$ as resultant of all forces that undergoes and whose external surface is generally crossable by matter.

But in this way (74) appears inevitably approximate if P substitutes V only because this has negligible dimensions as sufficiently small with respect to those that have importance in the given context.

Therefore we believe this approximate character of (74) and each its error of description of the material world must be eliminated adjoining, inasmuch moreover completely obvious, the further clarification of being P the center of mass of V.

In these considerations regarding (74), we also can consider P as an infinitesimal volume that has $\mathbf{W}$, contains M and undergoes $\mathbf{F}$, but in this way we introduce a physical object obviously inexistent in the material world and that is not definable as limit.

The quantities M and $M\mathbf{W}$ are named mass and momentum of P, but only abstractly because a point cannot contain mass. However, analogously to how (44) has been deduced by $\delta M = \rho\delta V$, by $\delta M\mathbf{W} = \rho\mathbf{w}\delta V$ and the being $\delta M\mathbf{W}$ the momentum of c we deduce that the momentum of C is $\iiint_V \rho\mathbf{w}\,\mathrm{d}V$, resulting, from this, Æ$\langle \mathfrak{C}, \mathfrak{v} \,/\!/\, C, V \rangle$ and (72), that $\mathfrak{M}\mathbf{W}_{\mathfrak{v}}$ is the momentum really possessed by $\mathfrak{C}$.

The (74) includes the law of conservation of momentum, inasmuch affirms that are constant the momenta of a material point and of a body that have null the respective resultants of all forces.

The originating formulation of the Newton's second law mentions neither particular points of application of vectors nor the concept of material point (successively introduced by Euler as it is reported in [6] and [88]). We believe such formulation, using current names and concepts, can be "the resultant of all forces undergone by a volume is equal to the temporal derivative of its momentum", that we express

$$\mathbf{F} = \frac{\mathrm{d}\iiint_V \rho\mathbf{w}\,\mathrm{d}V}{\mathrm{d}t}$$

where $\mathbf{F}$ is the resultant of all forces undergone by V.

Indeed the meaning we have attributed to (74) is substantially based on neglecting differences among parts of V, and doing the same with such expression, we have

$$\mathbf{F} = \frac{\mathrm{d}\iiint_V \rho\mathbf{w}\,\mathrm{d}V}{\mathrm{d}t} = \frac{\mathrm{d}\iiint_V \rho\mathbf{W}\,\mathrm{d}V}{\mathrm{d}t} = \frac{\mathrm{d}M\mathbf{W}}{\mathrm{d}t}$$

that includes the (74) itself.

7.2 Cauchy's equation of motion and the New Law

The Æ⟨$\mathfrak{m}\mathbf{w}_{\mathfrak{v}}$ // $M\mathbf{w}$ // (74)⟩ implies $\mathrm{d}\mathfrak{m}\mathbf{w}_{\mathfrak{v}}/\mathrm{dt} = \mathbf{F}_{\mathrm{T}}$ of which (73) and (64). This and the genericity of $\mathfrak{v}$ that consents to eliminate the integrals give rise to

$$\rho\left(\frac{\partial \mathbf{w}}{\partial t} + \nabla\mathbf{w}\cdot\mathbf{w}\right) = \mathbf{f}_{\mathrm{BE}} - \nabla P - \nabla\cdot\boldsymbol{\tau}_{\mathrm{D}} \tag{75}$$

whose ρ, $\mathbf{w}$, $\mathbf{f}_{\mathrm{BE}}$, P and $\boldsymbol{\tau}_{\mathrm{D}}$ are functions of $\underline{ж}$, and that, for Æ⟨$\mathbf{w}$ // f // (35)⟩ and (66), also has the form

$$\rho\frac{\mathrm{d}\mathbf{w}(\underline{x}, t)}{\mathrm{dt}} = \mathbf{f}_{\mathrm{BE}} - \nabla\cdot\boldsymbol{\tau}_{\mathrm{T}}$$

known as Cauchy's equation of motion.

For some authors (e.g. [27, 30, 31, 80]) the Newton's second law expressed by (74) is considered valid only if M is constant. However this regards not the fundamental (75) inasmuch its previous deduction is not modified by such diminution of applicability of (74).

From (66) we have

$$\frac{\mathbf{f}_{\mathrm{BE}} - \nabla P - \nabla\cdot\boldsymbol{\tau}_{\mathrm{D}}}{\rho} = \boldsymbol{f}_{\text{Ћ}} - \frac{\nabla P}{\rho} = \boldsymbol{f}_{\mathrm{T}} \tag{76}$$

of which

$$\boldsymbol{f}_{\mathrm{T}} = \frac{\mathbf{f}_{\mathrm{T}}}{\rho}, \qquad \boldsymbol{f}_{\text{Ћ}} = \frac{\mathbf{f}_{\text{Ћ}}}{\rho},$$

and where

$$\boldsymbol{f}_{\text{Ћ}}, \quad -\frac{\nabla P}{\rho} \text{ (based to (63) and Æ⟨}\boldsymbol{\tau}_{\mathrm{P}}, \delta\mathrm{v} \,//\, \boldsymbol{\tau}, \delta\mathbf{v} \,//\, (55)\text{⟩)} \quad \text{and} \quad \boldsymbol{f}_{\mathrm{T}}$$

are respectively the resultants per mass unit of all forces undergone by $\delta\mathfrak{v}$, $\delta\mathrm{v}$, and $\delta\mathrm{v}\cup\delta\mathfrak{v}$. Multiplying for ρ^{-1} or ρ a quantity per unit of volume (e.g. $\mathbf{f}_{\mathrm{T}}$) or mass (e.g. $\boldsymbol{f}_{\mathrm{T}}$) we obtain the same but per unit of mass or volume.

The Æ⟨$\mathbf{w}$ // f // (34)⟩ and (76) of which Æ⟨$\boldsymbol{f}_{\text{Ћ}}$ // f // (31)⟩ allow to write (75) as

$$\frac{\partial \mathbf{w}(\underline{ж})}{\partial t} + \frac{\mathrm{d}\mathbf{w}_{c}(\underline{x})}{\mathrm{dt}} = \boldsymbol{f}_{\text{Ћ}c}(\underline{x}) - \frac{\nabla P(\underline{ж})}{\rho(\underline{ж})} \tag{77}$$

The limit of (74) as $\mathsf{V} \to 0$, in the case $\mathrm{d}\mathsf{M}/\mathrm{dt} = 0$ that we have if V is the volume of a body, gives rise, on the basis of

$$\lim_{\mathsf{V}\to 0}(\mathsf{M}, \boldsymbol{F}) = (\delta\mathfrak{m}, \delta\mathfrak{m}\boldsymbol{f}_{\text{Тс}}),$$

to

$$\frac{\mathrm{d}\mathbf{w}_c(\underline{x})}{\mathrm{dt}} = \boldsymbol{f}_{\text{Тс}}(\underline{x}) \qquad (78)$$

This affirms the equality between the acceleration of an infinitesimal corpuscle and the resultant per mass unit of all forces that it undergoes. In literature this concept, strictly linked to the Newton's second law, also intervenes as application of such law (e.g. (4.3) in [5], section 2.1.2 in [7], (1.93) in [47]) or as immediately evident (e.g. (2.1) in [85]).

From (77) and (78) follows

$$\rho(\underline{ж})\frac{\partial\mathbf{w}(\underline{ж})}{\partial t} + \nabla\mathrm{P}(\underline{ж}) = 0 \qquad (79)$$

which is the new and independent law announced by the title of this paper.

The falsity of (79), and so also the presence of errors in the argumentation whereby it has been deduced, it could be proven by at least a case of validity of the $\rho\partial\mathbf{w}/\partial t \neq -\nabla P$ whose $P \equiv \mathrm{P}$ was ascertain by the being $\{P, \rho, \mathrm{T}\}$ the three variables of the equation of state.

The (77), (78) and (79) are coherent with the inherence of

$$\left\{\frac{\mathrm{d}\mathbf{w}(\underline{x}(t), t)}{\mathrm{dt}}, \boldsymbol{f}_{\mathrm{T}}\right\}, \quad \left\{\frac{\mathrm{d}\mathbf{w}_c(\underline{x}(t))}{\mathrm{dt}}, \boldsymbol{f}_{\text{Тс}}\right\} \quad \text{and} \quad \left\{\frac{\partial\mathbf{w}(\underline{ж})}{\partial t}, -\frac{\nabla\mathrm{P}}{\rho}\right\}$$

to the respective $\delta\mathrm{v} \cup \delta\mathfrak{v}$, $\delta\mathfrak{v}$ and $\delta\mathrm{v}$, as we deduce from

$$Æ\langle\mathbf{w} \,/\!/\, f \,/\!/\, (34), (35)\rangle \qquad \text{and} \qquad (76)$$

in coherence with the incompatibility among $Æ\langle\partial\mathbf{w}/\partial t \,/\!/\, f \,/\!/\, (31)\rangle$ and constancy of the $\underline{x}$ implied by each $\partial f(\underline{ж})/\partial t$.

Chapter 8

Models of Continuum Mechanics and New Law

8.1 The general model and its best known specifications

A flux is a quantity that crosses, possibly by transforming i.e. changing its qualities, a surface in a time range; so is a quantity exchanged, i.e. respectively delivered/lost and received/acquired, by two volumes separated by such a surface and is identifiable as a same (in absolute value) variation own of both. In adding the fluxes of a volume, i.e. the quantities that it exchanges with its surroundings, we use the convention of considering respectively positive or negative a quantity by it acquired or lost. In coherence with this, a flux can also in particular be defined as an applied vector (e.g. $f\boldsymbol{w}_S$ and $\boldsymbol{\phi}_F$ of section 3.2).

The first law of Thermodynamics is a balance equation for the energy of a volume and usually is formulated equaling the sum of its fluxes (e.g. heat and work are fluxes of energy respectively thermal and mechanical), plus its internal generation namely the quantity that it acquires or loses inasmuch is place of chemical and/or nuclear reactions, to the variation of the total energy that it contain. However generally neither the total energy nor its variation have expressions known and independent by such equation, and so is unknown also the total energy $\mathcal{E}_T$ (per unit of mass or volume) of a given infinitesimal volume (like $\delta\mathfrak{v}$, $\delta\mathrm{v}$ and $\delta\mathrm{v} \cup \delta\mathfrak{v}$).

More precisely, according to the usual literature, such an expression can be obtained by $\mathcal{E}_T$ sum of the kinetic and internal energy, and by a further constitutive equation i.e. the caloric equation of state. But to this regard we not prosecute with further considerations, because the first law of Thermodynamics is outside of this work and we believe we can treat it later more specifically.

The fundamental general model of the Continuum Mechanics (MCG) is a system of PDEs constituted by expressions of general laws valid for every substance (balance or conservation laws) and by equations peculiarly own of the given substance (constitutive equations): continuity equation i.e. first of (45); Cauchy's equation of motion i.e. (75); equation of state (that links P, ρ, T); stress constitutive equation (that links $\boldsymbol{\tau}_{\mathrm{D}}$ to tensorial variables that describe the deformation, having the $\tau_{\mathrm{D}ij} = \tau_{\mathrm{D}ji}$ that is deduced from Æ$\langle\tau_{\mathrm{D}}$ // τ // section 6.2.2$\rangle$ and is equivalent to a balance of angular momentum with respect to a fixed point); expressions of deformation tensors as functions of displacements and velocities such as those of $\mathbf{e}$, $\boldsymbol{\varepsilon}$, $\dot{\mathbf{e}}$ and $\dot{\boldsymbol{\varepsilon}}$ in section 5; first law of Thermodynamics adapted to Continuum Mechanics (i.e. balance equation for the energy of an infinitesimal volume) and (if available) caloric equation of state.

In the twelve scalar equations of such model, that (as we place implicit) remain having not available the caloric equation of state and after having applied the substitution method to eliminating the deformation tensors and their expressions, we have thirteen unknown functions of ж that express the respective unknowns ρ, w_1, w_2, w_3, P, T, $\tau_{\mathrm{D}11}$, $\tau_{\mathrm{D}22}$, $\tau_{\mathrm{D}33}$, $\tau_{\mathrm{D}12}$, $\tau_{\mathrm{D}13}$, $\tau_{\mathrm{D}23}$, $\mathcal{E}_{\mathrm{T}}$, also being possible the presence of other unknowns in the stress constitutive equation (as the λ of (68)).

A solution of a PDEs system, numerical as is implicit inasmuch not knowable that exact constituted by its unknown functions, consists of values of these and, with particular reference to [15] and [1], is calculable only if the number of such functions is not greater of the number of independent equations, in which case the system is said solvable.

This implies that a solution of MCG, which is constituted by twelve independent scalar equations in at least thirteen unknown functions, we can only compute it by reducing enough the number of unknown functions and/or replacing at least one of them with a known expression that adequately approximates it and contains no further unknowns.

Therefore we imply the absences of the balance equation for energy justified by the unavailability of a such expression of $\mathcal{E}_{\mathrm{T}}$ and of the equation of state justified by being the only where appear T and by not having interest to this variable. Indeed these absences not increase the difference between the numbers of functions unknown and equations that become respectively eleven and ten.

With reference to section 5, a stress constitutive equation, that we indicate

$$\boldsymbol{\tau}_D = \boldsymbol{\tau}_D(\mathsf{D}, \underline{e}) \qquad \text{of which} \qquad \mathsf{D} \equiv \{\mathbf{e} \vee \boldsymbol{\varepsilon}, \dot{\mathbf{e}} \vee \dot{\boldsymbol{\varepsilon}}\}$$

(implying that also it may be an implicit form $f(\boldsymbol{\tau}_D, \mathsf{D}, \underline{e}) = 0$), is inevitably approximate because determined by means of models sub-macroscopic and/or experiments, as the other constitutive relations which characterize a given substance, like the equation of state or the expressions of specific heats. But a $\boldsymbol{\tau}_D = \boldsymbol{\tau}_D(\mathsf{D}, \underline{e})$ is affected by errors of greater importance inasmuch caused by inadequate description of the deformation provided by $\mathbf{e}$, $\boldsymbol{\varepsilon}$ and $\dot{\boldsymbol{\varepsilon}}$, by the being $\boldsymbol{\tau}_D$ defined in relation to $\delta\mathfrak{v}$ as cube when however only an infinitesimal deformation can be represented by a such $\delta\mathfrak{v}$, and in every case by the being the function $\boldsymbol{\tau}_D = \boldsymbol{\tau}_D(\mathsf{D}, \underline{e})$ unknown and difficult to approximate because of peculiar difficulties of identification and measurement of the variables.

Thus MCG, as other drawbacks further to the having at least one unknown function too many and to the consequent necessity of physical approximations to achieve to an equal number of functions unknown and equations, is also affected by the errors particularly relevant typical of a stress constitutive equation.

Accepting as sufficiently small the error of approximate the equation of state with a relation among the only P and ρ (barotropicity), MCG becomes solvable if in such its eleven equations (the twelve said initially minus the balance equation for the energy) not appear unknowns further to the as many P, ρ, w_1, w_2, w_3, τ_{D11}, τ_{D22}, τ_{D33}, τ_{D12}, τ_{D13}, τ_{D23}.

Having not interest to distinguish $\mathsf{P}\boldsymbol{\delta}$ as the part of $\boldsymbol{\tau}_T$ indicated by (65) and having

$$\nabla \mathsf{P} + \nabla \cdot \boldsymbol{\tau}_D = \nabla \cdot \boldsymbol{\tau}_T \tag{80}$$

highlighted by (66), (75) and $\boldsymbol{\tau}_D = \boldsymbol{\tau}_D(\mathsf{D}, \underline{e})$ (whose D is implicitly clarified as function of $\mathbf{w}$) become

$$\rho\left(\frac{\partial \mathbf{w}}{\partial t} + \nabla \mathbf{w} \cdot \mathbf{w}\right) = \mathbf{f}_{BE} - \nabla \cdot \boldsymbol{\tau}_T \qquad \boldsymbol{\tau}_T = \boldsymbol{\tau}_T(\mathsf{D}, \underline{e}) \tag{81}$$

whereby MCG reduce to a model (MPM) solvable if in its ten equations not appear unknowns further to the as many ρ, w_1, w_2, w_3, τ_{T11}, τ_{T22}, τ_{T33}, τ_{T12}, τ_{T13}, τ_{T23}.

A specification of MCG, very well known for its great generality in describing fluid motions, consists in the Navier-Stokes equations the are obtained adopting as stress constitutive equation the first of (68). Sometimes is so named the only (75) whose $\boldsymbol{\tau}_D$ is expressed by such adoption. Introducing (70) in the scalar form

$$\rho\left(\frac{\partial w_i}{\partial t}+\sum_{j=1}^{3}\frac{\partial w_i}{\partial x_j}w_j\right)=f_{BEi}-\frac{\partial P}{\partial x_i}-\sum_{j=1}^{3}\frac{\partial \tau_{Dji}}{\partial x_j}$$

of (75) where $\mathbf{f}_{BE}\equiv f_{BEi}$, we obtain

$$\rho\left(\frac{\partial w_i}{\partial t}+\sum_{j=1}^{3}\frac{\partial w_i}{\partial x_j}w_j-f_{BEi}\right)+\frac{\partial P}{\partial x_i}+\mu\sum_{j=1}^{3}\sum_{h=1}^{3}\sum_{k=1}^{3}\delta_{jikh}\frac{\partial^2 w_k}{\partial x_h\partial x_j}=0$$

of which

$$f_{BEi}\equiv \boldsymbol{f}_{BE}=\frac{\mathbf{f}_{BE}}{\rho}$$

and that is a form (particularly convenient to computerize) of (75) subjected to the second of (68).

The Navier-Stokes equations are famous for the difficulties of calculating their solutions also only numerical, so much so that the achievement of a progress of their comprehension is one of the seven *Millennium Problems* (*https://www.claymath.org/millennium-problems*) of the *Clay Mathematics Institute*. In this regard we believe the peculiarly hard difficulties in conveniently solving such equations applied to real cases are essentially caused by the (68) that moreover are poorly improvable because of aforementioned errors peculiarly implied by stress constitutive equations.

We name perfect or ideal (inasmuch strictly speaking inexistent in the material world) a fluid that verifies

$$\{\tau_{Tij}=0,\tau_{Tii}=\tau_{Tjj};\forall i\neq j\} \tag{82}$$

from which, in coherence with (65), we deduce

$$\begin{gathered}\{\tau_{Dij}=\tau_{Tij}=0,\tau_{Dii}=\tau_{Djj};\forall i\neq j\}\\ \boldsymbol{\tau}_T=\mathbf{0}\qquad \boldsymbol{\tau}_T=\overline{P}\boldsymbol{\delta}\qquad \overline{P}=\tau_{Tii}=\tau_{Dii}+P\end{gathered} \tag{83}$$

In relation to such (82) and (83) we have not place the instead usually immediate $\overline{\mathsf{P}} \equiv \mathsf{P}$, because this, moreover to not being immediately evident, is, how we will see, incompatible with (79).

By referring MCG to such a fluid we have the following Euler equations, that, being freed from (68), simplify efficaciously those of Navier-Stokes in numerous technical approximations in which the viscous stresses are negligible e.g. the inviscid flow outside a boundary layer ([44]). Indeed (80) and $\nabla \cdot \boldsymbol{\tau}_{\mathrm{T}} = \nabla \overline{\mathsf{P}}$ (that is deduced from $\boldsymbol{\tau}_{\mathrm{T}} = \overline{\mathsf{P}}\boldsymbol{\delta}$ like the $\nabla \cdot \boldsymbol{\tau}_{\mathrm{P}} = \nabla \mathsf{P}$ of (63) has been deduced from $\boldsymbol{\tau}_{\mathrm{P}} = \mathsf{P}\boldsymbol{\delta}$), entail, as specification of MCG, the Euler equations consisting in the first of (45) and

$$\rho\left(\frac{\partial \mathbf{w}}{\partial t} + \nabla \mathbf{w} \cdot \mathbf{w}\right) = \mathbf{f}_{\mathrm{BE}} - \nabla \overline{\mathsf{P}}$$

of which $\overline{\mathsf{P}} = \tau_{\mathrm{Tii}}$ and that, being a PDEs system of four scalar equations in the five unknowns ρ, w_1, w_2, w_3 and $\overline{\mathsf{P}}$, to become solvable must be integrated by a further independent equation devoid of unknowns different of the said five. Between the modes of achieve such solvability we have barotropicity and incompressibility, respectively consisting of a correlation among the $\{\overline{\mathsf{P}}, \rho\}$ and of the constancy of $\rho(\underline{ж})$ with ρ a known characteristic of the given substance.

The Euler equations for an incompressible fluid (MEI) are deduced by replacing, in those aforementioned, the first of (45) with the $\nabla \cdot \mathbf{w} = 0$ which is deduced from it and constancy of $\rho(\underline{ж})$, and obtaining thus a PDEs system solvable inasmuch constituted by four scalar equations in the as many unknowns w_1, w_2, w_3, $\overline{\mathsf{P}}$.

The scalar product of $\mathbf{w}_{c}$ and vectorial equation of MEI, stationarity (i.e. $\{\partial f(\underline{ж})/\partial t = 0; \forall f\}$) and Æ$\langle \mathbf{w} \,/\!/\, f \,/\!/\, (34)\rangle$ give rise to

$$\rho \mathbf{w}_{c} \cdot \frac{d\mathbf{w}_{c}}{dt} = \mathbf{w}_{c} \cdot \mathbf{f}_{\mathrm{BE}} - \mathbf{w}_{c} \cdot \nabla \overline{\mathsf{P}}$$

This; $ff' = (f^2/2)'$ and Æ$\langle \mathbf{w}_{c} \,/\!/\, \mathbf{a} \,/\!/\, (5)\rangle$ (regardind first addend); body forces consisting of the only conventional standard gravity (i.e. $\mathbf{f}_{\mathrm{BE}} = -g\hat{\mathbf{e}}_3$ of which $g = 9.80665$ m/s^2 and $\hat{\mathbf{e}}_3$ upward directed), $d\mathbf{x}_{c}/dt \equiv \mathbf{w}_{c}(\underline{x})$ (in (33)), (32), $\mathbf{x}_{c} \equiv x_{\mathrm{i}}$ and (19); give rise to

$$\frac{1}{2}\frac{d|\mathbf{w}_{c}|^2}{dt} + g\frac{dx_3}{dt} + \frac{1}{\rho}\frac{d\overline{\mathsf{P}}_{c}}{dt} = 0$$

from whose indefinite integral with respect to t we have the Bernoulli's equation (MEB)

$$\frac{|\mathbf{w}_c|^2}{2} + g x_3 + \frac{\overline{P}_c}{\rho} = \mathcal{K}_B$$

fundamental in hydraulics because its erroneity in the modeling the material world (due to the approximations from which has been deduced) can be measured by the variation of $\mathcal{K}_B$ (i.e. of the sum in the first member) along the trajectory $\mathcal{T}$ of c.

An incompressible and immobile ($\mathbf{w} = \mathbf{0}$) fluid verifies (82) and so $\nabla \cdot \boldsymbol{\tau}_T = \nabla \overline{P}$ (of which $\overline{P} = \tau_{Tii}$). This, (80) and $\boldsymbol{f}_{BE} = -g\hat{\boldsymbol{e}}_3$ imply that for such a fluid MCG reduces to the hydrostatic MIS

$$\nabla \overline{P} = -\rho g \hat{\boldsymbol{e}}_3 \tag{84}$$

which is exactly solvable because equivalent to the

$$\frac{\partial \overline{P}}{\partial x_1} = \frac{\partial \overline{P}}{\partial x_2} = 0 \qquad \text{and} \qquad \frac{\partial \overline{P}}{\partial x_3} = -\rho g$$

that, used in a curvilinear integral, from $\underline{ж}_0$ of which $\underline{ж}_0 \equiv (x_{01} x_{02}, x_{03}, t)$ to $\underline{ж}$, of the expression of $d\overline{P}$ that we have from Æ$\langle \overline{P} \,//\, f \,//\, (19) \rangle$, give rise to

$$\overline{P}(\underline{ж}) = \overline{P}(\underline{ж}_0) + \rho g (x_{03} - x_3)$$

The (84) of which $\mathbf{w} = \mathbf{0}$ shows $\overline{P} \equiv P$ incompatible with (79) that affirms ∇P related to $\partial \mathbf{w}(\underline{ж})/\partial t$. However $\overline{P} \neq P$ (i.e. $\overline{P} \not\equiv P$) in MEI and MIS can be deduced as follows. The (69) and $\nabla \cdot \mathbf{w} = 0$, that respectively follow from the first of the (68) and from incompressibility and/or $\mathbf{w} = \mathbf{0}$, give rise to $\overline{P} = P$. This and (83) imply $\boldsymbol{\tau}_D = \mathbf{0}$ and so indeformability that however in MEI and MIS is unmotivated inasmuch not implicated by incompressibility while instead this is implicated by the other (a deformation can happen incompressibly i.e. by varying shape but not extent, and a compression cannot happen indeformably). This can constitute a *demonstratio per absurdum* of having to exclude, from this two models, (68) and (69), and therefore also of being able to deduce the $\overline{P} \neq P$ on the basis of (83) and $\boldsymbol{\tau}_D \neq \mathbf{0}$.

But in this respect is relevant both that $\overline{P} = P$ could result from significant measures of $\overline{P}$, ρ and T that show these three variables as the only of an equation which consequently would necessarily be that of state where $\overline{\mathsf{P}} \equiv \mathsf{P}$, and that *vice versa* $\overline{\mathsf{P}} \neq \mathsf{P}$ could result from experiments showing variations of $\overline{\mathsf{P}}$ correspondent to constancy of ρ and T.

In case of stationarity, incompressibility and $\boldsymbol{f}_{\mathrm{BE}} = -g\hat{\boldsymbol{e}}_3$, the Navier-Stokes equations become the following fundamental stationary model of a Newtonian, isotropic and incompressible fluid (MNS)

$$\nabla \cdot \mathbf{w} = 0 \qquad \rho \nabla \mathbf{w} \cdot \mathbf{w} + \nabla \cdot \boldsymbol{\tau}_{\mathrm{D}} + \nabla \mathsf{P} + \rho g \hat{\boldsymbol{e}}_3 = 0$$

$$\boldsymbol{\tau}_{\mathrm{D}} = \mu\left(\nabla \mathbf{w} + (\nabla \mathbf{w})^{\mathrm{T}}\right) \tag{85}$$

which is solvable because constituted by ten scalar equations in the as many unknowns w_1, w_2, w_3, P, $\tau_{\mathrm{D}11}$, $\tau_{\mathrm{D}22}$, $\tau_{\mathrm{D}33}$, $\tau_{\mathrm{D}12}$, $\tau_{\mathrm{D}13}$, $\tau_{\mathrm{D}23}$. In the [51, 52, 53, 54] are described applications, of the program PEEI ([1]) presented and available (freeware) in *https://www.giacomo.lorenzoni.name/peei/*, to calculate the solutions of four fluid motions (Hagen-Poiseuille, Taylor-Couette laminar, Couette plane, Poiseuille plane) modeled by specifications of MNS.

The discordance of (84) of which $\overline{\mathsf{P}} \neq \mathsf{P}$ from the $\nabla \mathsf{P} = -g\hat{\boldsymbol{e}}_3$, that would follow from MNS and $|\mathbf{w}| = 0$, it is eliminated by admitting that MNS and in particular (85) are worth only for $|\mathbf{w}| > 0$.

The $\mathbf{w} = \mathbf{0}$, incompressibility, $\mathbf{f}_{\mathrm{BE}} = -\rho g\hat{\boldsymbol{e}}_3$ and adoption of the first of (67) as stress constitutive equation that specifies the second of (81), transform MPM in the elastostatic linear (MEL)

$$\nabla \cdot \boldsymbol{\tau}_{\mathrm{T}} + \rho g \hat{\boldsymbol{e}}_3 = 0 \qquad \boldsymbol{\tau}_{\mathrm{T}} = \mu\left(\nabla \mathbf{s} + (\nabla \mathbf{s})^{\mathrm{T}}\right) + \lambda(\nabla \cdot \mathbf{s})\delta$$

that is solvable inasmuch of nine scalar equations in the as many unknowns $\tau_{\mathrm{T}11}$, $\tau_{\mathrm{T}22}$, $\tau_{\mathrm{T}33}$, $\tau_{\mathrm{T}12}$, $\tau_{\mathrm{T}13}$, $\tau_{\mathrm{T}23}$, s_1, s_2, s_3, and that, in the area of important disciplines such as Mechanical Component Design and Building Science, has the very wide and fundamental use explained, in comparison with the errors generally induced by stress constitutive equations and abstractness of incompressibility, by the typical smallness of the treated displacements and by the absence of such errors in the case of an infinitesimal $|\mathbf{s}|$. In [55, 56, 57, 58] are described applications of said program PEEI to calculate the solution of four elastostatic cases (elastic torsion of a circular bar, hydrostatic compression of a elastic sphere, pure elastic bending of a prismatic bar, elastic axial extension of a prismatic rod) modeled by specifications of MEL.

8.2 Influence of the New Law and a new general model

In absence of (79), MIS, MEL, MNS and MEB are not subjected to the condition

$$\text{“}\mathsf{P}(\underline{ж}) \quad \text{and} \quad \mathsf{T}(\underline{ж}) \quad \text{both} \quad \text{constant”} \tag{86}$$

because their stationarity and incompressibility do not impede the occurrence of $\nabla\mathsf{P} \neq \mathbf{0}$ and $\nabla\mathsf{T} \neq \mathbf{0}$ complying with the equation of state.

Instead, since this equation must be in force even if absent in the given model, and following $\nabla\mathsf{P} = \mathbf{0}$ from (79) and stationarity, we have

$$\left\{(79), \quad \left(\frac{\partial f(\underline{ж})}{\partial \mathsf{t}} = 0; \forall f\right), \quad \rho(\underline{ж}) \text{ constant}\right\} \quad \Rightarrow \quad (86)$$

whereby (79) implies that MIS, MEL, MNS and MEB are subjected to $\mathsf{P}(\underline{ж})$ constant and also to isothermy ($\mathsf{T}(\underline{ж})$ constant).

However this not seems constitute a decisive incongruence, because, also if such four models do not explicitly exclude $\nabla\mathsf{T} \neq \mathbf{0}$, though normally concern contexts in which T and $\nabla\mathsf{T}$ are devoid of interest, relevance and applicative feedbacks, and so MIS, MEL, MNS and MEB are *de facto* implicitly integrated by the isothermy that (79) imposes them. Moreover, alternatively to such isothermy, (79) and $\nabla\mathsf{T} \neq \mathbf{0}$ become compatible if it is possible to admit that, also with $\mathsf{P}(\underline{ж})$ constant, the given gradients of T and the equation of state do not entail gradients of ρ large enough to compromise the incompressibility approximately own of the four models in object.

From (78) and

$$Æ\langle \mathbf{w}, \boldsymbol{f}_{\text{ћ}} \,/\!/\, f \,/\!/\, (31), (34)\rangle$$

we deduce $\nabla\mathbf{w}\cdot\mathbf{w} = \boldsymbol{f}_{\text{ћ}}$. This, (76) and $\boldsymbol{f}_{\mathrm{BE}} = \mathbf{f}_{\mathrm{BE}}/\rho$ give rise to

$$\boldsymbol{f}_{\mathrm{BE}} - \frac{\nabla\cdot\boldsymbol{\tau}_{\mathrm{D}}}{\rho} = \nabla\mathbf{w}\cdot\mathbf{w} \tag{87}$$

A system of equations has the same solutions of another deduced from it by substituting an equation with a linear combination of this and other of the same system. Hence, being (75) a linear combination of (79) and (87), the system, obtainable by adding the New Law (79) to MCG, has the same solutions as the one that is distinguished as NMG and is deduced by eliminating (75) and adding (79) and (87).

The same MIS, MEL, MNS and MEB are also deducible from NMG, as from MCG but with the difference of the (86) imposed as said by (79).

The solvability of NMG is no worse than that of MCG, since it has three more scalar equations and same unknown functions, even if though the calculability of solutions of NMG should be checked on a case-by-case basis, because the having at least so many independent equations in as many unknowns is a condition necessary but not sufficient.

To this regard we believe however of preeminent importance that the comparison between MCG and NMG if anyway of minor interest, since both are affected by the error of modeling the material world caused by the said more relevant errors introduced by stress constitutive equations.

We note instead more interesting the model *MOD* that is obtained by eliminating, from NMG, (87) and the stress constitutive equation, and that so is composed by first of the (45), (79), equation of state and balance equation for the energy. The so having also eliminated $\boldsymbol{f}_{\mathrm{BE}}$ together with (87) is not worrying since a function, that analogously to $\boldsymbol{f}_{\mathrm{BE}}$ expresses the body forces, is certainly present in the balance equation for the energy.

Indeed such *MOD* has six scalar equations where appear the only seven unknown functions that express the respective $\mathsf{\rho}$, w_1, w_2, w_3, P, T, $\mathcal{E}_{\mathrm{T}}$, and hence, in addition to the decisive value of being not affected by said peculiar errors caused by the stress constitutive equations, also has the one that its solvability has as necessary condition the substitution of an only unknown function with an its adequate known approximation.

In this regard we anticipate that a next work will relate to the theory exposed in the [77, 78, 79, 81] and in truth extremely innovative for various aspects in addition to the decisive result recognizable in a balance equation for the energy of general validity, independent and where we can eliminate both $\mathcal{E}_{\mathrm{T}}$ and each unknown additional to the other six of *MOD*.

This *MOD*, devoid of errors of modeling the material world different from those typical (ineliminable but usually reducible) of every constitutive equation that describes a certain substance, and mathematically solvable without the necessity of approximations of laws valid for every substance inasmuch has six equations in as many unknowns, will be able to constitute a general thermomechanical representation yet absent in the Continuum Mechanics and whose peculiar importance also philosophical follows by the being this the science of the common sensory reality.

In MCG and its said specifications has not been explicitly introduced the second law of Thermodynamics, that in Continuum Mechanics is represented by the Clausius-Duhem inequality and that though can be in the constitutive relations. In the *MOD* this law is present as an expression of the heat of friction as a function of velocity, and moreover results realistic and conform to such law are also related to the realism of the additional conditions (like those initial or boundary) always necessary for the solution of a PDEs system.

Conclusion

Believing satisfactorily achieved the aim of this work that is a reliable theorization of the validity of the New Law (79), now we aim to define analogous argumentations for the other innovative laws of the *MOD*, that, as mentioned in the introduction and section 8.2, are essentially a balance equation for the energy from which we can eliminate the further unknown function that expresses the total energy (inasmuch we establish an its expression that do not contains other unknowns in addition to those already present) and an expression of the heat generated by friction.

Indeed, completed such a second step, it remains the activity, almost unlimited but quickly satisfactory in the absence of unfavorable results, of verifying the truthfulness of the *MOD* (i.e. its capacity of representing the real material world) solving it for the innumerable possible applications and being able in this regard already to rely on the abundantly validated solver PEEI ([1]) of PDEs.

References

[1] G. Lorenzoni, *PEEI: a Computer Program for the Numerical Solution of Systems of Partial Differential Equations*, https://www.giacomo.lorenzoni.name/peei/, July 2019.

[2] H. Altenbach, *Fundamentals of continuum mechanics – classical approaches and new trends*, J. Phys.: Conf. Ser. 991 012003, April 2018, https://doi.org/10.1088/1742-6596/991/1/012003.

[3] G. Lorenzoni, *The True Probability of a Confidence Interval*, November 2018, Aracne, Roma.

[4] D. Gray, W. Huebsch, *The balance principle and the Reynolds transport theorem in introductory fluid mechanics*, August 2017, https://doi.org/10.1177/0306419017726629.

[5] X. Oliver, C. A. de Saracibar, *Continuum Mechanics for Engineers. Theory and Problems*, 2nd ed., March 2017, https://doi.org/10.13140/RG.2.2.25821.20961.

[6] M. Stan, *Euler, Newton, and Foundations for Mechanics*, The Oxford Handbook of Newton, 2017, https://doi.org/10.1093/oxfordhb/9780199930418.013.31.

[7] J. Málek, V. Průša, *Derivation of Equations for Continuum Mechanics and Thermodynamics of Fluids*, Handbook of Mathematical Analysis in Mechanics of Viscous Fluids, editors Giga & Novotný, Springer, 2016, https://doi.org/10.1007/978-3-319-10151-4_1-1.

[8] K. R. Rajagopal, *On the Flows of Fluids Defined through Implicit Constitutive Relations between the Stress and the Symmetric Part of the Velocity Gradient*, Vol. 1, Fluids 2016, https://doi.org/10.3390/fluids1020005.

[9] S. Zuccher, *Note di Fluidodinamica*, Università degli Studi di Verona, 2016, http://profs.sci.univr.it/~zuccher/downloads/fd-zuccher.pdf.

[10] G. Buresti, *A note on Stokes' hypothesis*, Acta Mech 226, 3555–3559 (2015), https://doi.org/10.1007/s00707-015-1380-9.

[11] R. Mauri, *Thermodynamics and Evolution*, In: *Transport Phenomena in Multiphase Flows*, Fluid Mechanics and Its Applications, vol 112. Springer Cham 2015, https://doi.org/10.1007/978-3-319-15793-1_1.

[12] R. Abeyaratne, *Continuum Mechanics*, Volume II of Lecture Notes on The Mechanics of Elastic Solids, Department of Mechanical Engineering, 77 Massachusetts Institute of Technology Cambridge, MA 02139-4307, USA, 2015, http://web.mit.edu/abeyaratne/lecturenotes.html.

[13] T. Ruggeri, *New Frontiers In Non-Equilibrium Thermodynamics*, Conference: Frontiere - Accademia Nazionale dei Lincei - Atti dei Convegni Lincei - Bardi Edizione, Vol. 314, Roma 2015.

[14] A. Bertram, R. Glüge, *Solid Mechanics - Theory, Modeling, and Problems*, Springer International Publishing Switzerland 2015, https://doi.org/10.1007/978-3-319-19566-7.

[15] G. Lorenzoni, *A method to numerically solve every differential analytical model*, Bollettino di Matematica pura e applicata dell'Università degli Studi di Palermo, Vol. VIII, dicembre 2015.

[16] D. Kleppner, R. Kolenkow, *An Introduction to Mechanics*, 2nd ed., Cambridge University Press, 2014, http://www.cambridge.org/9780521198110.

[17] F. H. Koenemann, *Cauchy's stress theory in a modern light*, Eur. J. Phys. 35 (2014) 015010 (15pp), https://doi.org/10.1088/0143-0807/35/1/015010.

[18] W. Chen, *The renaissance of continuum mechanics*, Journal of Zhejiang University-SCIENCE A (Applied Physics & Engineering), Zhejiang University and Springer-Verlag Berlin Heidelberg 2014, https://doi.org/10.1631/jzus.A1400079.

[19] A. Sharma, *Isaac Newton, Leonhard Euler and $F = ma$*, Physics Essays, Volume 27, Number 3, September 2014, pp. 503-509(7), https://doi.org/10.4006/0836-1398-27.3.503.

[20] A. Tangredi, G. Frosali, *Meccanica dei Continui*, Dipartimento di Matematica e Informatica U. Dini, Università degli Studi di Firenze, Firenze 2014.

[21] F. Costa, G. Peruzzi, *I Principia di Newton - Le basi della dinamica classica*, Dipartimento di Fisica e Astronomia "Galileo Galilei", Università degli Studi di Padova, aprile 2014.

[22] G. Lorenzoni, *Argomentazioni analitiche di probabilità e statistica*, Aracne, Roma 2013.

[23] N. Phan-Thien, *Tensor Notation*, In: *Understanding Viscoelasticity. Graduate Texts in Physics*, Springer-Verlag Berlin Heidelberg 2013, https://doi.org/10.1007/978-3-642-32958-6_1.

[24] A. Tamir, *Conservation Law of Mass*, J. Chem. Eng. Process. Technol., Vol. 4, I. 8, 2013, https://doi.org/10.4172/2157-7048.1000e114.

[25] G. Guzzetta, *Relating Deformation and Thermodynamics: An Opportunity for Rethinking Basic Concepts of Continuum Mechanics*, Entropy 2013, 15, 2548-2569, https://doi.org/10.3390/e15072548.

[26] D. Violeau, *Fluid Mechanics and the SPH Method*, Oxford University Press, 2012.

[27] B. Samardžija, S. Šegan, *Movement of a Body With Variable Mass*, Publ. Astron. Obs. Belgrade No. 91 (2012), 97 - 104.

[28] E. Tadmor, R. Miller, R. Elliott, *Continuum Mechanics and Thermodynamics From Fundamental Concepts to Governing Equations*, Cambridge University Press 2012, http://www.cambridge.org/9781107008267.

[29] V. QUINN, A. STUBBLEFIELD, *Continuum and Solid Mechanics (Concepts and Applications)*, Academic Studio, Delhi 2012.

[30] Y. ICHIKAWA, A. P. S. SELVADURAI, *Introduction to Continuum Mechanics*, In: *Transport Phenomena in Porous Media*, Springer Berlin Heidelberg 2012, https://doi.org/10.1007/978-3-642-25333-1_2.

[31] D. GROSS, W. HAUGER, J. SCHRÖDER, W. A. WALL, S. GOVINDJEE, *Engineering Mechanics 3 - Dynamics*, Springer-Verlag, Berlin Heidelberg 2011, https://doi.org/10.1007/978-3-642-14019-8.

[32] J. W. RUDNICKI, *Fundamentals of Continuum Mechanics*, Department of Civil and Environmental Engineering and Department of Mechanical Engineering, Northwestern University, Evanston, IL, 2011.

[33] Y. I. DIMITRIENKO, *Nonlinear Continuum Mechanics and Large Inelastic Deformations*, Springer Science+Business Media B.V. 2011, https://doi.org/10.1007/978-94-007-0034-5.

[34] Z. MARTINEC, *Continuum Mechanics (Lecture Notes)*, Department of Geophysics, Faculty of Mathematics and Physics, Charles University in Prague, 2011.

[35] H. WONG, C. J. LEO, N. DUFOUR, *Thermodynamics in Mono and Biphasic Continuum Mechanics*, 2011, https://doi.org/10.5772/20235.

[36] B. POURCIAU, *Is Newton's second law really Newton's?*, American Journal of Physics 79, 1015 (2011), https://doi.org/10.1119/1.3607433.

[37] A. BORRELLI, *Meccanica dei Continui*, Dipartimento di Matematica, Università degli Studi di Ferrara, 2011.

[38] A. RUINA, R. PRATAP, *Introduction to Statics and Dynamics*, Oxford University Press (Preprint), 2010.

[39] M. E. GURTIN, E. FRIED, L. ANAND, *The Mechanics and Thermodynamics of Continua*, Cambridge University Press, 2010, http://www.cambridge.org/9780521405980.

[40] J. R. RICE, *Solid Mechanics*, School of Engineering and Applied Sciences, Harvard University, Cambridge USA 2010, http://esag.harvard.edu/rice/e0_Solid_Mechanics_94_10.pdf.

[41] W. M. LAI, E. KREMPL, D. RUBIN, *Introduction to Continuum Mechanics*, 4th ed., Elsevier Inc. 2010.

[42] J. L. WEGNER, J. B. HADDOW, *Elements Of Continuum Mechanics And Thermodynamics*, Cambridge University Press 2009, http://www.cambridge.org/9780521866323.

[43] V. Y. FORTOV, *Equation of State*, Thermopedia, 11 February 2011, https://doi.org/10.1615/AtoZ.e.equation_of_state.

[44] V. M. EPIFANOV, *Boundary Layer*, Thermopedia, 16 March 2011, https://doi.org/10.1615/AtoZ.b.boundary_layer.

[45] S. NAIR, *Introduction To Continuum Mechanics*, Cambridge University Press 2009, http://www.cambridge.org/9780521875622.

[46] F. IRGENS, *Continuum Mechanics*, Springer-Verlag, Berlin Heidelberg 2008.

[47] A. A. SHABANA, *Computational Continuum Mechanics*, Cambridge University Press 2008, http://www.cambridge.org/9780521885690.

[48] L.H. SÖDERHOLM, *Basic Continuum Mechanics*, Department of Mechanics, KTH, S-100 44 Stockholm, Sweden, Stockholm 2008.

[49] O. GONZALEZ, A. M. STUART, *A First Course in Continuum Mechanics*, Cambridge University Press 2008, http://www.cambridge.org/9780521886802.

[50] J. N. REDDY, *An Introduction to Continuum Mechanics*, Cambridge University Press 2008, http://www.cambridge.org/9780521870443.

[51] G. LORENZONI, *Hagen-Poiseuille Flow*, PEEI's application, 16/11/2008, https://doi.org/10.5281/zenodo.2530110.

[52] G. LORENZONI, *Laminar Taylor-Couette Flow*, PEEI's application, 14/11/2008, https://doi.org/10.5281/zenodo.2296323.

[53] G. Lorenzoni, *Plane Couette Flow*, PEEI's application, 28/10/2008, https://doi.org/10.5281/zenodo.2530134.

[54] G. Lorenzoni, *Plane Poiseuille Flow*, PEEI's application, 26/10/2008, https://doi.org/10.5281/zenodo.2645347.

[55] G. Lorenzoni, *Elastic Torsion of a Circular Bar*, PEEI's application, 16/09/2008, https://doi.org/10.5281/zenodo.3229590.

[56] G. Lorenzoni, *Hydrostatic Compression of a Elastic Sphere*, PEEI's application, 16/09/2008, https://doi.org/10.5281/zenodo.3068165.

[57] G. Lorenzoni, *Pure Elastic Bending of a Prismatic Bar*, PEEI's application, 05/09/2008, https://doi.org/10.5281/zenodo.2759475.

[58] G. Lorenzoni, *Elastic Axial Extension of a Prismatic Rod*, PEEI's application, 03/09/2008, https://doi.org/10.5281/zenodo.2759467.

[59] L. A. Segel, G. H. Handelman, *Mathematics Applied to Continuum Mechanics*, Society for Industrial and Applied Mathematics, Philadelphia 2007.

[60] G. Querzoli, *Dispense di Meccanica dei Fluidi*, Dipartimento di Ingegneria del Territorio, Facoltà di Ingegneria, Università degli Studi di Cagliari, 2006.

[61] A. Romano, R. Lancellotta, A. Marasco, *Continuum Mechanics using Mathematica®. Fundamentals, Applications and Scientific Computing*, Birkhäuser, Boston 2006.

[62] H. C. Wu, *Continuum Mechanics and Plasticity*, Chapman & Hall/CRC Press, Boca Raton 2005.

[63] C. Man, R. L. Fosdick, *The Rational Spirit in Modern Continuum Mechanics*, Essays and Papers Dedicated to the Memory of Clifford Ambrose Truesdell III, Springer Science + Business Media Inc. 2005.

[64] D. Zaccaria, *Concetti della Meccanica del Continuo*, Dipartimento di Ingegneria Civile, Università di Trieste, 2005.

[65] D. Tong, *Classical Dynamics*, University of Cambridge Part II Mathematical Tripos, Michaelmas Term, 2004 and 2005, http://www.damtp.cam.ac.uk/user/tong/dynamics.html.

[66] M. Jirásek, *Nonlocal Theories in Continuum Mechanics*, Acta Polytechnica Vol. 44 No. 5–6/2004, Czech Technical University Publishing House, https://ojs.cvut.cz/ojs/index.php/ap/article/view/610.

[67] R. Temam, A. Miranville, *Mathematical Modeling in Continuum Mechanics*, 2nd ed., Cambridge University Press 2005, http://www.cambridge.org/9780521617239.

[68] J. C. Kolecki, *An Introduction to Tensors for Students of Physics and Engineering*, NASA/TM—2002-211716, National Aeronautics and Space Administration, Glenn Research Center, Cleveland, Ohio, 2002.

[69] T. Hopman, *Introduction to indicial notation*, Department of Physics, University of Guelph, 2002.

[70] I. V. Kazachkov, V. A. Kalion, *Numerical Continuum Mechanics*, Vol. 1, Div. of Heat and Power Technology, Department of Energy Technology, The Royal Institute of Technology, Stockholm, Sweden, 2002.

[71] P. M. Naghdi, *P. M. Naghdi's Notes on Continuum Mechanics*, Department of Mechanical Engineering, University of California at Berkeley, ME 185, 2001.

[72] G.T. Mase, G.E. Mase, *Continuum Mechanics for Engineers*, 2nd edition, CRC Press, Boca Raton 1999.

[73] V. E. Saouma, *Introduction to Continuum Mechanics and Elements of Elasticity/Structural Mechanics*, Dept. of Civil Environmental and Architectural Engineering, University of Colorado 1998.

[74] J. Bonet, R. D. Wood, *Nonlinear Continuum Mechanics For Finite Element Analysis*, Cambridge University Press 1997.

[75] J.H. Heinbockel, *Introduction to Tensor Calculus and Continuum Mechanics*, Department of Mathematics and Statistics Old Dominion University, 1996.

[76] M. Gad-el-Hak, *Stokes' Hypothesis for a Newtonian, Isotropic Fluid*, Journal of Fluids Engineering, Vol. 117, pp. 3–5, March 1995, https://doi.org/10.1115/1.2816816.

[77] G. Lorenzoni, *Una definizione di procedimenti numerici nella nuova termodinamica di equilibrio (NTE)*, Tecnica Italiana n.2, 1995, G. Zorzut Cormons (Gorizia).

[78] G. Lorenzoni, *Una conferma numerica di una nuova termodinamica di equilibrio (NTE)*, Tecnica Italiana n.2, 1994, G. Zorzut Cormons (Gorizia).

[79] G. Lorenzoni, *Un modello termomeccanico di applicabilità generale basato su una nuova formulazione del primo principio della termodinamica*, Tecnica Italiana n.1, 1993, G. Zorzut Cormons (Gorizia).

[80] A. R. Plastino, J. C. Muzzio, *On the use and abuse of Newton's second law for variable mass problems*, Celestial Mech Dyn Astr 53, 227–232 (1992), https://doi.org/10.1007/BF00052611.

[81] G. Lorenzoni, *È proposta la possibilità di nuove formulazioni quantitative del primo e secondo principio*, Atti del 47-mo Congresso Nazionale della Associazione Termotecnica Italiana, Parma 15-18 settembre 1992.

[82] C. A. Truesdell, *A First Course in Rational Continuum Mechanics*, 2nd ed., Academic Press Inc., 1991.

[83] A. Rutherford, *Vectors, Tensors, and the Basic Equations of Fluid Mechanics*, Dover Publications Inc., New York 1989.

[84] D. C. Kay, *Tensor Calculus*, McGraw-Hill Companies Inc., New York 1988.

[85] L.D. Landau, E.M. Lifshitz, *Fluid Mechanics*, 2nd ed., Pergamon Press, Oxford 1987.

[86] M. E. GURTIN, *An Introduction to Continuum Mechanics*, Academic Press Inc., New York 1981.

[87] L.E. MALVERN, *Introduction to the Mechanics of a Continuous Medium*, Prentice-Hall Inc, Englewood Cliffs 1969.

[88] C. TRUESDELL, *Essays in the History of Mechanics*, Springer-Verlag, Berlin Heidelberg 1968.

[89] M. E. GURTIN, W. O. WILLIAMS, *An axiomatic foundation for continuum thermodynamics*, Department of Mathematical Sciences, Carnegie Mellon University, 1967.

[90] H. SEMAT, R. KATZ, *Physics, Chapter 11: Rotational Motion (The Dynamics of a Rigid Body)*, University of Nebraska - Lincoln, Research Papers in Physics and Astronomy, Robert Katz Publications, 1958, `http://digitalcommons.unl.edu/physicskatz/141`.

[91] H. S. M. COXETER, *Regular Polytopes*, Methuen & Co. Ltd., London, 1948.

[92] I. NEWTON, *Philosophiae Naturalis Principia Mathematica (Mathematical Principles Of Natural Philosophy)*, translated by Andrew Motte, originally published in latin in 1687, published in english in 1728, edition created and published by Global Grey 2015.

www.ingramcontent.com/pod-product-compliance
Lightning Source LLC
Chambersburg PA
CBHW080948170526
45158CB00008B/2418
9781008945494